Palgrave Studies in Green Criminology

Criminologists have increasingly become involved and interested in environmental issues to the extent that the term Green Criminology is now recognised as a distinct subgenre of criminology. Within this unique area of scholarly activity, researchers consider not just harms to the environment, but also the links between green crimes and other forms of crime, including organised crime's movement into the illegal trade in wildlife or the links between domestic animal abuse and spousal abuse and more serious forms of offending such as serial killing. This series will provide a forum for new works and new ideas in green criminology for both academics and practitioners working in the field, with two primary aims: to provide contemporary theoretical and practice-based analysis of green criminology and environmental issues relating to the development of and enforcement of environmental laws, environmental criminality, policy relating to environmental harms and harms committed against non-human animals and situating environmental harms within the context of wider social harms; and to explore and debate new contemporary issues in green criminology including ecological, environmental and species justice concerns and the better integration of a green criminological approach within mainstream criminal justice. The series will reflect the range and depth of high-quality research and scholarship in this burgeoning area, combining contributions from established scholars wishing to explore new topics and recent entrants who are breaking new ground.

More information about this series at
https://link.springer.com/bookseries/14622

Mark Wright

Responding to Environmental Crimes

Lessons from New Zealand

palgrave
macmillan

Mark Wright
Faculty of Law
University of Canterbury
Christchurch, New Zealand

Palgrave Studies in Green Criminology
ISBN 978-3-030-89249-4 ISBN 978-3-030-89250-0 (eBook)
https://doi.org/10.1007/978-3-030-89250-0

Cover credit: georgeclerk

This Palgrave Macmillan imprint is published by the registered company Springer Nature Switzerland AG
The registered company address is: Gewerbestrasse 11, 6330 Cham, Switzerland

Preface

On 16 May 2016 I was sitting in the Tauranga District Court, as junior counsel for the Bay of Plenty Regional Council, listening to Mobil Oil New Zealand Limited being sentenced for discharging a contaminant into water, in contravention of section 15(1)(a) of New Zealand's Resource Management Act 1991 (**RMA**) (*Bay of Plenty Regional Council v Mobil Oil New Zealand Ltd* [2016] NZDC 8903). About a year earlier, a pipeline that Mobil used to refuel ships at Mount Maunganui Port had developed two holes and spilled, "most likely", 3,000–4,000 litres of Heavy Fuel Oil into Tauranga harbour (at [16]). The Judge adopted, as a starting point (the sentence considered objectively appropriate for the offending), a fine of $360,000, which was primarily based on the culpability of the defendant and the harm caused. In particular, the discharge occurred because Mobil had failed to repair a previously identified issue with the pipeline, an omission which was described by the Judge as involving "recklessness bordering on deliberateness" (at [35]), with the spill causing serious and widespread damage to a harbour that is rich in cultural heritage and popular for fishing, diving and swimming (at [19]).

The starting point fine was increased by one-third to take into account the wealth of the defendant—Mobil had a net income of NZ$60 million for the period ending 31 December 2014 and its parent company had earnings of US$16.2 billion in 2015 (*Bay of Plenty Regional Council v Mobil Oil New Zealand Limited—Agreed Summary of Facts*, available at www.boprc.govt.nz)—and then reduced by 20% for personal mitigating factors. These including paying out direct costs of NZ$1.8 million to the Council and other parties to remediate the damage caused, and for committing "to a process with local iwi to try and address its relationship and take a role as a good corporate citizen within the Tauranga Moana" (at [50]–[55]). The sentence was then reduced by a further 25% for an early guilty plea.

The end sentence was a conviction and a fine of $288,000, which was (and still is) one of the highest monetary penalties ever imposed under the RMA. Yet the Judge concluded his decision by lamenting that the Council's "actual costs in relation to the prosecution… [had]… been over $300,000 yet it only recovers 90% of the fine" (at [59]), and so:

> [60] It appears to me that there may need to be some review of the legislation in due course to take into account that these costs are otherwise borne by the rate payers [sic] through no cause of their own.

This comment was the genesis of this book. The Judge appeared to be suggesting that the penalty imposed on a defendant in a successful RMA prosecution should fully reimburse the local authority that initiated the matter (section 342 of the Act provides that, in the usual course, 90% of any fines imposed are to be paid to the prosecuting council). However, while the "polluter pays" principle is well-established in environmental law, compensation, and in particular to the prosecutor for its costs incurred in bringing a case, is not traditionally a priority of the courts when it comes to criminal sentencing. Equally, it could have been said that the legislation should be reviewed because even a fine of $288,000 would be unlikely to deter other members of the regulated community, given how rarely RMA prosecutions occur. For instance, in the 2015/2016 financial year, when this offending occurred, there were just 57 prosecutions commenced, compared to 10,985 resource consent breaches and 31,446 complaints (an unknown number of which were substantiated), not to mention all of the breaches of the Act that would have occurred but been undetected. It would also have been unlikely to have deterred the defendant itself, given both its earnings and those of its parent company (as referred to above). On the other hand, one could have argued that there is nothing wrong with the legislation, as it had enabled the imposition of a "deserved" sentence—that is, one that is proportionate to the seriousness of the offence committed and which takes into account aggravating and mitigating factors personal to the defendant.

In truth, having previously prosecuted for the Crown, where the more typical fare are crimes such as assault and burglary, the RMA's offences and penalties, and the institutional arrangements for prosecuting, had already struck me as strange. The offences are very broad, both in terms of the *actus rei* and the lack of any requirement that *mens rea* be proved, meaning that local authorities have great discretion as to who to charge and for what. Further, the penalties that can be imposed include both fines and imprisonment (which means that alternatives such as supervision, intensive supervision, community work, community detention and home detention are also options), yet sentencing almost always regresses to what dollar amount the offender should pay, with such amounts being insurable. The fact that guilty defendants receive a conviction, which is usually considered a penalty in and of itself, appears almost irrelevant, except for those rare prosecutions where the defendant applies for a discharge without conviction, in which case both sides argue strongly why a conviction is, or is not, suddenly of such importance. When cases come to Court they are not classified as either private or public prosecutions, and local authorities are represented by a range of different advocates, including in-house counsel, barristers, large law firms and Crown Solicitors' offices, with the type of representation (and who bears the cost of it) sometimes changing partway through the matter. Finally, cases are presided over by judges who spend most of their time in civil proceedings, and are managed accordingly.

I would spend the next five years looking at the sentencing of RMA offenders and the Act's offences, penalties and institutional arrangements for prosecuting, as well as everything in between. My goal was to answer the deceptively simple question of whether the approach the RMA takes to environmental crime is working, and the results are recorded in this book. As I write this, these results have taken on an added impetus, with the New Zealand Government having recently committed to replacing the RMA, meaning that now is the time to learn from the previous endeavours (or, as it transpires, mistakes). But environmental crime is a global phenomenon, and there are also lessons in this book for those beyond New Zealand's shores.

There are a number of people and institutions that have made this book possible. I have been fortunate to receive guidance from many of my colleagues at the University of Canterbury, including John Hopkins, Elizabeth Toomey, Neil Boister and James Mehigan. I also received funding for my research from the University of Canterbury and the New Zealand Law Foundation, which was of much assistance. Thanks to Rob White from the University of Tasmania, who encouraged me to submit the proposal that turned into this book and who provided advice along the way. Thanks also to my research assistant, Selma Kafedžić, who provided useful comments on the manuscript and who helped me put the finishing touches on it. My final note of thanks is reserved for my family—my parents Rosemary and Murray, my brother Matt and his wife and children, and my wife Nikki and our three boys—Oscar, Albie and Jude. Thank you for your love and support.

Christchurch, New Zealand Mark Wright
September 2021

Contents

List of Figures

Part I

Setting the Scene

1
Introduction

Environmental harm is a worldwide problem, and New Zealand is no exception. Despite the images portrayed on the Tourism New Zealand website of lush green countryside, snow-capped mountains and clear blue rivers and lakes, together with the accompanying slogan "100% Pure New Zealand" (Tourism New Zealand 2021), scratch beneath the surface and all is not quite so rosy. A 2021 report on land use found that "nationwide 80 percent of monitoring sites failed to meet the targets for at least one soil quality indicator" (Ministry for the Environment and Stats NZ 2021, 17) and "the area of land covered with native ecosystems continue to shrink, mainly through conversion to agriculture or forestry" (Ministry for the Environment and Stats NZ 2021, 34). A 2020 report on freshwater found, among other things, that native species and ecosystems are under threat and water is polluted in urban, farming and forestry areas (Ministry for the Environment and Stats NZ 2020b). Another report that same year, which this time focussed on the atmosphere and climate, noted that New Zealand's gross carbon dioxide emissions were 7.7 tonnes per person in 2017—17th out of 32 OECD countries (Ministry for the Environment and Stats NZ 2020a). The list goes on.

M. Wright, *Responding to Environmental Crimes*, Palgrave Studies in Green Criminology, https://doi.org/10.1007/978-3-030-89250-0_1

One way that New Zealand attempts to deal with environmental harm is via the Resource Management Act 1991 (RMA). This Act, New Zealand's primary environmental regulation statute, has as its key themes "the sustainable management of natural and physical resources, the integrated management of resources, and the control of the adverse effects of activities on the environment" (Palmer 2015, 3.12). Importantly for present purposes, the Act sets out a range of duties and restrictions relating to the use of land, subdivision, the coastal marine area, the beds of rivers and lakes, the taking and use of water and discharges of contaminants, which the 78 local authorities throughout the country—a mixture of territorial authorities, regional councils and unitary authorities—are responsible for enforcing. To secure compliance with these duties and restrictions, the local authorities are provided with various tools, including abatement notices, enforcement orders and (criminal) offences. The latter are the ultimate sanction that can be imposed for non-compliance, with natural persons being liable to imprisonment for a term not exceeding 2 years or a fine not exceeding $300,000, and non-natural persons being liable to a fine not exceeding $600,000, and the courts also have the ability to make an enforcement order or an order that the offender's resource consent be reviewed (RMA, section 339).

There have been some investigations into compliance, monitoring and enforcement of the RMA generally. In 2011, a report by the Auditor-General raised concerns about variable practice between local authorities when using the enforcement provisions, and the involvement of elected members in specific prosecution decisions (Office of the Auditor-General 2011). Momentum gathered on this front over the following decade, with reports by the Ministry for the Environment and the Environmental Defence Society (the latter a not-for-profit organisation) that highlighted, among other things, limited resourcing of compliance, monitoring and enforcement, inconsistent enforcement practices throughout the country and a lack of data (Ministry for the Environment 2016; Brown 2017). This has culminated in a finding by the Resource Management Review Panel, appointed in 2019 to undertake a "comprehensive review of the resource management system" (2020, 8), that there is weak compliance, monitoring and enforcement by local authorities (2020, 21).

Studies have also been undertaken on the use of restorative justice under the RMA. In particular, New Zealand has been described as "providing an exemplary model of the innovative use of conferencing" (Al-Alosi and Hamilton 2019, 1462), with section 24A of the Sentencing Act 2002, which applies to offending across-the-board, dictating that consideration of restorative justice is mandatory where the offender has pleaded guilty to the offence and there are one or more victims. However, it must be observed that the uptake of restorative justice under the RMA has been relatively limited, with Hamilton noting that conferences were held in just 4% of environmental offending prosecutions between 30 June 2002 and 30 September 2012, and only in 16 instances in the period from 1 October 2012 to 1 July 2020 (2021, 130–132).

What has been missing is any sustained analysis of the Act's offences, penalties and institutional arrangements for prosecuting *in and of themselves*. This has meant that the responses to the issues identified, such as the Ministry for the Environment's *Best Practice Guidelines for Compliance, Monitoring and Enforcement under the Resource Management Act 1991* (2018) and the granting of powers to the Environmental Protection Authority (a central government agency), pursuant to the Resource Management Amendment Act 2020, to start investigations, assist local authorities with enforcement and even intervene in existing matters, have focussed on compliance, monitoring and enforcement generally. It also means that, while much is now known about restorative justice under the Act, there is little awareness of what happens in those cases—the vast majority—where, for whatever reason, restorative justice is not undertaken.

This book contains a case study on the offences in the Act, with the offences assessed by reference to two criteria—are they effective and are they being used by the local authorities appropriately. The conclusion reached is that the offences are falling down on both fronts, and this is due to local authorities treating prosecutions more like they are civil proceedings. The ultimate argument advanced is that local authorities act in this way because the offences, penalties and institutional arrangements for prosecuting, while formally criminal, are in substance more akin to those found in civil matters.

There are three lessons that can be taken from the case study. First, it shows the need to align the substance of offences, penalties and institutional arrangements with their criminal form. Second, it highlights how collecting good data on the use of offences requires an understanding as to how the criminal law works. Third and finally, it raises questions about the ability of the criminal law *in general* to deter.

Introducing the Case Study

As just mentioned, the RMA's offences are assessed by reference to two criteria, namely whether they are effective and whether they are being used by the local authorities appropriately. While the offences and the framework for assessment are discussed more fully in Chapters 2 and 3 (respectively), it is important to briefly explain these two concepts of effectiveness and appropriateness upfront. This is the task of this section, before the case study is placed in the field of green criminology (in the following section).

Effectiveness is in itself uncontroversial, but difficult in application. More precisely, while everyone would agree that offences should achieve their purpose, the RMA does not provide any guidance as to what this purpose should be, and there are differing views in the literature as to the proper rationale for imposing penalties. In particular, is it to deter would-be offenders, as regulatory scholars typically suggest (see, for instance, Abbot 2009), or is it because the sanction is "deserved", as criminal law scholars favour (see, for instance, Von Hirsch 1993)? This book does not take a normative position on this question, but rather the effectiveness of the offences is critiqued against why the local authorities and the courts say they are prosecuting and imposing sanctions (respectively).

On the other hand, appropriateness is a more divisive criterion, because it depends on how one views the criminal law. As Simester and Von Hirsch note:

> ...some writers see the criminal law as just another state regulatory tool, one that happens to involve fines and imprisonment rather than injunctions and damages. To them, there is no fundamental distinction between crimes and other branches of law that impose sanctions. (2011, 4)

According to this approach, which is common among those that write about environmental non-compliance (although rarely articulated as such), "the criminal law exists to visit punishments such as imprisonment or fines in situations where other forms of disincentive make for an insufficient deterrent" (Simester and Von Hirsch 2011, 4). The question of whether offences are being used appropriately is, therefore, of little (if any) concern (but see, as an exception, Lees 2015).

Yet criminal law scholars, while differing among themselves as to precisely when the criminal law should be employed, generally reject such a view. The criminal law, they argue, has a "social significance" not found with other means of imposing penalties (Simester and Von Hirsch 2011, 4). When the legislature employs the criminal law this means that those tasked with enforcement are being committed to doing so in a way that, among other things, treats those subject to it equally and is fair. This normative position is adopted here, with the use of the offences by local authorities critiqued against values such as these.

Two final points on the case study need to be made at this time. First, it must be noted that the Resource Management Review Panel (referred to earlier) has recommended that the RMA be repealed and replaced (2020, 23), and this has been accepted by the New Zealand Government (Parker 2021). However, the way in which environmental non-compliance should be dealt with will remain a live issue in any new legislation, and as will be seen mistakes have been made with the RMA that need to be avoided in the future. In addition, the question of whether offences should be used to support an environmental regulation statute, and if so how, is not by any means unique to New Zealand. As White and Heckenberg have noted, "[d]etailed exploration of one jurisdiction can provide useful concepts and benchmarks by which to compare environmental laws cross-nationally" (2014, 257), and so by investigating the New Zealand situation it is hoped that other jurisdictions can also gain some learnings.

Second, it is important to note that this investigation into the RMA's offences is done with full awareness that non-compliance with the RMA is not the only, or even necessarily the main, reason for the environmental issues New Zealand faces. Indeed, as noted in the following section, one of the key things green criminology has drawn attention to is environmental damage that is caused legally—the "lawful but awful" (White and Heckenberg 2014, 3). But that is not a reason to eschew the study of environmental non-compliance. Rather, it is suggested, understanding how best to respond to unlawful conduct will put us in a better position to deal with other (currently lawful) environmental harms should they subsequently be proscribed.

The Case Study and Green Criminology

Green criminology is an umbrella term (Hall 2016; Nurse 2016), coined in 1990 (Lynch 1990), that refers to:

> …the study by criminologists of environmental harm (that may incorporate wider definitions of crime than are provided by strictly legal definitions); environmental laws (including enforcement, prosecution and sentencing practices); and environmental regulation (systems of criminal, civil and administrative law designed to manage, protect and preserve specified environments and species, and to manage the negative consequences of particular industrial processes). (White and Heckenberg 2014, 8)

It is, accordingly, wide-ranging, and this is particularly the case when one considers that the term "environmental", while undefined (Gibbs et al. 2010, 126), is typically interpreted broadly. For example, White has noted that there are "brown" environmental issues, such as pollution, oil spills and disposal of toxic/hazardous waste, "green" environmental issues, such as habitat destruction, loss of wildlife and depletion of the ozone layer, and "white" environmental issues, such as genetically modified organisms, food irradiation and environmentally related communicable diseases (2005, 278; for further definitions of green criminology and green crimes, see Lynch et al. 2017, 9). There is also often a transnational element to these issues, with "many such transgressions transcend[ing] the normal boundaries of legality, jurisdiction, geography and social divide" (White and Heckenberg 2014, 25).

An important aspect of green criminology is the concept of environmental crime (White and Heckenberg 2014, 8). The reason for this is captured by Skinnider, who has observed that environmental crimes:

> … often fail to prompt the required response by governments, the enforcement community and the public. Often perceived as "victimless", environmental crimes do not always produce an immediate consequence, the harm may be diffused or go undetected for a lengthy period of time. Added to this is the fact that many environmental disruptions are actually legal and take place with the consent of society. Classifying what is an environmental crime involves a complex balancing of communities' interest in jobs and income with ecosystem maintenance, biodiversity and sustainability. (2011, 1–2)

One of the important debates therefore, as highlighted in this quote, is how to define an environmental crime. As Situ and Emmons have observed, a "strict legalist perspective emphasizes that crime is whatever the criminal code says it is", whereas a "social legalist perspective argues that some acts, especially by corporations, may not violate the criminal law yet are so violent in their expression or harmful in their effects to merit definition as crimes" (Situ and Emmons 2000, 2–3; see also Clifford and Edwards 2012, Chapter 5; Lynch and Stretsky 2003). A legalistic definition is adopted in this book, as the focus is on the offences in the RMA, which naturally involves only what is currently unlawful. However, it is important to keep the wider definition in mind; for instance, when considering the reasons that local authorities act as they do when they detect non-compliance. By way of example, it may be that an enforcement officer is unwilling to sanction a more technical breach of the Act because the officer knows that other regulatees are committing more harmful, albeit lawful, acts.

White has noted that there are three main approaches to studying environmental crime (2008, 181). The first is a "socio-legal" approach, which focuses on the criminal law as it is and "[a]ttempts to improve [the] quality of investigation, law enforcement, prosecution and conviction on illegal environmentally-related activity" (2008, 182). The second is a "regulatory approach", which considers different ways in which social regulation can be used to "prevent and curtail environmental harm" (2008, 182). Third and finally, there is a "social action approach", which considers the "need for fundamental social change, and to challenge the hegemony of transnational capital and dominant nation-states in setting the environmental agenda" (2008, 182).

It will be apparent that this book fits most naturally into the first such approach. In particular, it seeks to "chart up existing environmental legislation and provide a sustained socio-legal analysis of specific breaches of law, the role of environmental law enforcement agencies, and the difficulties and opportunities of using criminal law against environmental offenders" (White 2013, 27). As already noted, it does so by way of a case study, a common approach used by green criminologists (Lynch et al. 2017, 29), which considers the effectiveness of the offences and whether they are being used appropriately.

The effectiveness assessment was primarily undertaken by reviewing Ministry for the Environment and Ministry of Justice data on, among other things, breaches of the RMA detected, prosecutions brought and sentences imposed for the period from the enactment of the RMA (1 October 1991) to 30 June 2017, with a particular focus on matters concluded during the 2016–2017 financial year. The precise nature of this data is referred to in Chapter 4, but it is important to recognise upfront that official information on environmental enforcement is notoriously patchy, and at times unreliable, and data on the use of the RMA's offences is no exception to this. This emphasises the importance of triangulation (White and Heckenberg 2014, 88), which was possible, at least to an extent, by looking at corroborating sources and making specific requests of councils pursuant to the Local Government Official Information and Meetings Act 1987.

The appropriateness assessment involved considering not only the manner in which the local authorities use the offences, but also their rationale for doing so. This was also problematic, as there was often scant information available regarding many of the decisions made and, where there was material available, there is always the suspicion that it has been "constructed more with a view to satisfying outsiders rather than accuracy" (Gill 2002, 542). The approach taken by Hawkins in his study of decision-making in the health and safety field was therefore followed, whereby he noted that by "employing different methods, and drawing from a variety of sources of data, problems arising from the elusiveness of decision-making can be mitigated to an extent" (2002, 450). This meant using a wide range of material, including council policies and reports on enforcement, national guidelines, studies of local authority enforcement behaviour (as part of which enforcement officers and other council officials had been interviewed) and case law.

Two final points on approach can be made. First, much like with the different definitions of environmental crime, aspects of the alternative approaches set out by White (regulatory and social action) remain relevant. For instance, the regulatory approach is necessary when it comes to understanding how local authorities use informal or directive (as opposed to sanctioning) tools in responding to environmental breaches, and the social action approach can help explain the role of the public and the (in)effectiveness with which the current regime deals with large corporate offenders. Second, parts of this book adopt an approach that could either be said to be at the "legal" extreme of a socio-legal approach or an outright legalistic approach (depending on the definitions adopted: compare Brisman and South 2019; White 2008). In particular, in order to understand the findings on effectiveness and appropriateness it was necessary to determine the legal constraints faced by courts when sentencing RMA offenders and what impact the drafting and interpretation of the Act's offences, penalties and institutional arrangements for prosecutions has had on the approach taken by local authorities. This was critical, given the ultimate aim of this book is to provide lessons, as:

> Greater familiarity with legal analysis and with legal conceptions of environmental harms and crimes could help criminologists and other social science scholars properly appraise existing legal and regulatory frameworks, and contribute more fully to the development of appropriate policy and legal strategies to respond to such harms and crimes. (Orlando and Bergin 2017, 7; see also Hall 2014, 106)

The legalistic approach is, though, taken one step further here, and the RMA's arrangements are compared to the philosophical basis for both sentencing and the employment of the criminal law as a response to environmental non-compliance. Such an approach has, it is suggested, been lacking in green criminology (but see, in terms of sentencing, O'Hear 2004; and aspects of criminal law theory are discussed in White and Heckenberg 2014, 9–13; Lin 2006), but was of importance here. This is because, as will be seen, it is suggested that the root of all problems with compliance, monitoring and enforcement of the RMA is the lack of connection between the offences, penalties and institutional arrangements for prosecuting and the normative foundations of the criminal law.

Outline of the Book

This book unfolds as follows. Part II (Chapters 2 and 3) sets up the framework for the case study, with Chapter 2 describing the RMA's offences, penalties and institutional arrangements for prosecuting, and placing these in context. The chapter begins with the RMA in general, observing that it is a regulatory statute that was enacted to correct for the presence of externalities. It then notes that the Act regulates in a command and control manner, with the "controls" split between directive (abatement notices and enforcement orders) and sanctioning (infringement offences and criminal offences) options. The RMA's offences, penalties and institutional arrangements for hearing prosecutions are described, before the key features of the Act are compared and contrasted to international models. The conclusion reached is that while much of the RMA is similar to overseas environmental regulation statutes, the biggest difference is New Zealand's institutional arrangements, with prosecutors not fitting neatly into either a public or private characterisation and the presence of specialist judges sitting in generalist courts. The other major divergence is that fines are returned to the enforcement agency rather than going into a central consolidated fund.

With Chapter 2 having identified two broad (and overlapping) theoretical areas that the RMA's offences fit into—regulatory enforcement and the criminal law—these areas are then the topic of Chapter 3. The chapter begins by noting that regulatory enforcement theory suggests that people comply (and do not comply) with rules for different reasons, and that this must be recognised in an enforcement strategy. It then goes through the various approaches that have been developed to secure obedience with the rules, observing that a "regulatory orthodoxy" has emerged that is broadly compliance-based. This promotes working with the regulatee rather than against them, and favours having a number of different enforcement tools and using sanctions only when required.

The focus then shifts to the different reasons why a sanction might be imposed, which is a key point of contention between regulatory enforcement and criminal law scholars. For the former, deterrence is typically preferred, with theory suggesting that people will be deterred where the expected cost of offending outweighs the expected benefit of doing so. Focussing on the expected cost of an offence, this is a function of the severity of the penalties imposed pursuant to the offence and the certainty that such a penalty will be imposed. On the other hand, criminal law scholars typically prefer desert theory, which suggests that penalties should be proportionate to the gravity of the offence committed.

The "criminal law" approach to imposing sanctions flows through to when offences should be applied to those in breach. Those responsible for enforcement have much discretion, and this must be applied, among other things, equally and fairly. The approach also has implications for the employment and drafting of offences. The observation is made that criminal law theory does not suggest that the criminal law is automatically inappropriate for regulatory breaches, as these can be public wrongs that are sufficiently harmful and committed culpably, but the manner in which offences in regulatory statutes are drafted is often overbroad.

The chapter concludes by setting out the framework for assessing the RMA's offences, which is whether they are effective and whether they are being used by the local authorities appropriately. In terms of effectiveness, given the availability of two viable options when considering the rationale for imposing sanctions, no normative position is taken on how to assess this. However, when it comes to appropriateness, it is argued that enforcers have been committed to using the offences in a way that complies with fundamental criminal law values. Conducting the assessment is the task of Part III (Chapters 4–7). Chapters 4 and 5 focus on the question of whether the offences are effective, first answering the question (Chapter 4) and then explaining why (Chapter 5). Chapters 6 and 7 focus on the question of whether the offences are being used appropriately, again first answering the question (Chapter 6) and then explaining why (Chapter 7).

As just noted, while the criminal law perspective is adopted for gauging whether the offences are being used appropriately, no normative position is taken on the question of effectiveness. Chapter 4 accordingly begins by considering why the local authorities say they are prosecuting RMA offenders and the courts say they are imposing penalties, and identifies that, for both of them, deterrence is the primary purpose. Effectiveness can therefore be assessed by gauging the expected cost of the penalties imposed pursuant to the offences, on the basis that the greater the expected cost the more of a deterrent (and so effective) the offences will be (and vice versa). In terms of severity, the usual penalty imposed for RMA offending is a modest fine (sometimes accompanied by an enforcement order aimed at securing compliance with the Act), with more punitive options rarely used and convictions and informal penalties doing little to bolster the "primary" sanction. This means that, overall, the severity of the penalties is no more than moderate. In terms of certainty, there is a low chance that a breach of the RMA will result in a criminal penalty, as many breaches are not detected, where a breach is detected formal enforcement action often does not follow, and if it does follow then it is rarely prosecution (although such prosecutions are usually successful). Given certainty is generally considered the more important of the two factors, this means that the offences appear to have a minimal expected cost, and so it would only take a correspondingly low expected benefit from breaching the RMA for a rational regulatee to offend. Accordingly, the offences would seem to be ineffective.

Why are criminal penalties only moderate and so rarely imposed? This is the subject of Chapter 5, and the question is considered from the perspective of the three parties that can influence the expected cost of the offences—local authorities (which can affect both the certainty that a penalty will be imposed and the severity of the penalty), the public (which can affect certainty only) and the courts (which can affect severity only). It transpires that it is a combination of constraints on the courts and the public, and choices made by local authorities, that is restricting the deterrent effect of the penalties. In terms of those whose actions are constrained, the courts follow orthodox, desert-based, sentencing practice, which has restricted fine levels due to the uncertainties around culpability, the low base point from which sentences started and the number of discounts that Judges feel obliged to give off the starting point. Similarly, the public does not, practically, have the means to increase the number of prosecutions or pressure local authorities to prosecute more. On the other hand, local authorities have choices in terms of both certainty and severity, and they apply limited resources to detecting breaches and are concerned about the cost, recovery and political palatability of prosecuting. This means they use the RMA's offences sparingly and, when they do prosecute, they typically seek compensatory, as opposed to more punitive, outcomes. This is, it is suggested, a distinctly "non-criminal" approach.

The combination of local authorities prosecuting sparingly and, when they do prosecute, seeking compensatory outcomes can be termed the "compensatory gloss", and Chapter 6 asks whether this way of using the offences is appropriate. It is shown that the compensatory gloss approach has resulted in councils rarely taking certain types of prosecutions (and favouring other types), entering into plea bargains with defendants to ensure that cases are resolved by non-natural persons pleading guilty (and agreeing on the summary of facts), and resolving cases out of court when they can achieve the same, if not better, outcomes as they would have achieved in a prosecution. However, in order to achieve these outcomes local authorities are, among other things, treating regulatees unequally and acting unfairly, in an unprincipled manner and even, potentially, unlawfully. Ultimately, the way local authorities are acting is, from a criminal law perspective, inappropriate, and more like a "civil" approach, yet they have been allowed to act in this way due to a lack of checks and accountability.

Chapter 7 finishes the critique of the RMA's offences by looking at why local authorities consider it acceptable to use the offences in this inappropriate way. It does this by analysing the drafting of the offences, penalties and institutional arrangements for prosecuting, and how they have been interpreted by the courts, to see whether such features are driving the approach taken by councils. It is observed that the offences are very broad, in terms of who can be prosecuted and for what, with the prosecution not needing to prove any mental element and the defendant only being able to escape liability if it can meet the restrictive statutory defences (that it must notify the prosecutor it is relying on). Further, the focus of the penalties is on insurable fines that are returned to the party bringing the charges, and convictions are typically said to not have any moral force. Finally, RMA prosecutions fit into neither of the usual categories of "private" or "public", local authorities can instruct whoever they want to act for them, and cases are managed by judges that do no other criminal work. Taken altogether, the offences and prosecutions are criminal in name only, and more akin to civil causes of actions and proceedings, meaning it is no surprise that local authorities apply them inappropriately (from a criminal law perspective).

Part IV, which consists of Chapter 8, concludes the book by setting out the three key lessons from the case study. First, the case study shows the importance of aligning the substance of offences, penalties and institutional arrangements with their criminal form. Second, it shows that collecting good data on the use of offences requires an understanding as to how the criminal law works. Third and finally, the case study also raises questions about the ability of the criminal law *in general* to deter. The book finishes by suggesting that there is still a role for the criminal law in condemning significant environmental wrongdoing, but any such offences, together with their penalties and institutional arrangements for prosecuting, should be drafted in accordance with criminal law norms to ensure they are treated with the respect they deserve.

References

Abbot, Carolyn. 2009. *Enforcing Pollution Control Regulation.* Portland: Hart Publishing.

Al-Alosi, Hadeel, and Mark Hamilton. 2019. The Ingredients of Success for Effective Restorative Justice Conferencing in an Environmental Offending Context. *University of New South Wales Law Journal* 42 (4): 1460–1488.

Brisman, Avi, and Nigel South. 2019. Green Criminology and Environmental Crimes and Harms. *Sociology compass* 13 (1). https://doi.org/10.1111/soc4.12650.

Brown, Marie. 2017. *Last Line of Defence: Compliance, Monitoring and Enforcement of New Zealand's Environmental Law*. Auckland: Environmental Defence Society.

Clifford, Mary, and Terry D. Edwards. 2012. *Environmental Crime*. Burlington: Jones & Bartlett Learning.

Gibbs, Carole, Meredith L. Gore, Edmund F. McGarrell, and I.I.I. Louie Rivers. 2010. Introducing Conservation Criminology: Towards Interdisciplinary Scholarship on Environmental Crimes and Risks. *British Journal of Criminology* 50 (1): 124–144. https://doi.org/10.1093/bjc/azp045.

Gill, Peter. 2002. Policing and Regulation: What Is the Difference? *Social & Legal Studies* 11 (4): 523–546. https://doi.org/10.1177/096466390201100403.

Hall, Matthew. 2014. The Roles and Use of Law in Green Criminology. *International Journal for Crime, Justice and Social Democracy* 3 (2): 96–109. https://doi.org/10.5204/ijcjsd.v3i2.176.

Hall, Matthew. 2016. Doing 'Green Criminology': Methodologies, Research Strategies and Values (or Lack Thereof?). In *Greening Criminology in the 21st Century: Contemporary Debates and Future Directions in the Study of Environmental Harm*, ed. Matthew Hall, Jennifer Maher, Angus Nurse, Gary Potter, Nigel South, and Tanya Wyatt, 25–41. Abingdon: Routledge.

Hamilton, Mark. 2021. *Environmental Crime and Restorative Justice: Justice as Meaningful Involvement*. Switzerland: Palgrave Macmillan.

Hawkins, Keith. 2002. *Law as Last Resort: Prosecution Decision-Making in a Regulatory Agency*. Oxford: Oxford University Press.

Lees, Emma. 2015. *Interpreting Environmental Offences: The Need for Certainty*. Oxford: Hart Publishing.

Lin, Albert. 2006. The Unifying Role of Harm in Environmental Law. *Wisconsin Law Review* 2006 (3): 897–985.

Lynch, Michael J. 1990. The Greening of Criminology: A Perspective on the 1990s. *Critical Criminologist* 2 (3): 1–4, 11–12.

Lynch, Michael J., Michael A. Long, Paul B. Stretesky, and Kimberly L. Barrett. 2017. *Green Criminology: Crime, Justice, and the Environment*. Oakland: University of California Press.

Lynch, Michael J., and Paul B. Stretsky. 2003. The Meaning of Green: Contrasting Criminological Perspectives. *Theoretical Criminology* 7 (2): 217–238. https://doi.org/10.1177/1362480603007002414.

Ministry for the Environment. 2016. Compliance, Monitoring and Enforcement by Local Authorities Under the Resource Management Act 1991.

Ministry for the Environment. 2018. Best Practice Guidelines for Compliance, Monitoring and Enforcement Under the Resource Management Act 1991.

Ministry for the Environment and Stats NZ. 2020a. New Zealand's Environmental Reporting Series: Our Atmosphere and Climate 2020.

Ministry for the Environment and Stats NZ. 2020b. New Zealand's Environmental Reporting Series: Our Freshwater 2020.

Ministry for the Environment and Stats NZ. 2021. New Zealand's Environmental Reporting Series: Our Land 2021.

Nurse, Angus. 2016. *An Introduction to Green Criminology and Environmental Justice*. London: Sage.

O'Hear, Michael. 2004. Sentencing the Green-Collar Offender: Punishment, Culpability, and Environmental Crime. *The Journal of Criminal Law and Criminology* 95 (1): 133–276.

Office of the Auditor-General. 2011. Managing Freshwater Quality: Challenges for Regional Councils.

Orlando, Emanuela, and Tiffany Bergin. 2017. Forging a Socio-Legal Approach to Environmental Harms. In *Forging a Socio-Legal Approach to Environmental Harms: Global Perspectives*, ed. Tiffany Bergin and Emanuela Orlando, 3–19. Abingdon: Routledge.

Palmer, Kenneth. 2015. Resource Management Act 1991. In *Environmental and Resource Management Law*, ed. Derek Nolan. Wellington: LexisNexis.

Parker, David. 2021. RMA to Be Repealed and Replaced. Press release.

Resource Management Review Panel. 2020. New Directions for Resource Management in New Zealand.

Simester, A.P., and Andreas Von Hirsch. 2011. *Crimes, Harms, and Wrongs: On the Principles of Criminalisation*. Oxford: Hart Publishing.

Situ, Yingyi, and David Emmons. 2000. *Environmental Crime: The Criminal Justice System's Role in Protecting the Environment*. Thousand Oaks: Sage.

Skinnider, Eileen. 2011. *Victims of Environmental Crime—Mapping the Issues.* Vancouver: The International Centre for Criminal Law Reform and Criminal Justice Policy.

Tourism New Zealand. 2021. Homepage. https://www.newzealand.com/int/. Accessed 17 May 2021.

Von Hirsch, Andreas. 1993. *Censure and Sanctions*. New York: Clarendon Press.

White, Rob. 2005. Environmental Crime in Global Context: Exploring the Theoretical and Empirical Complexities. *Current Issues in Criminal Justice* 16 (3): 271–285. https://doi.org/10.1080/10345329.2005.12036324.
White, Rob. 2008. *Crimes Against Nature: Environmental Criminology and Ecological Justice*. Devon: Willan Publishing.
White, Rob. 2013. The Conceptual Contours of Green Criminology. In *Emerging Issues in Green Criminology: Exploring Power, Justice and Harm*, ed. Reece Walters, Diane Westerhuis, and Tanya Wyatt, 17–33. Basingstoke: Palgrave Macmillan.
White, Rob, and Diane Heckenberg. 2014. *Green Criminology: An Introduction to the Study of Environmental Harm*. Abingdon: Routledge.

Part II

Establishing the Framework for Assessment

2

The RMA's Offences and Their Context

As White has noted, "[d]ifferent countries have different laws and frequently quite different approaches to dealing with environmental crime. It is important, therefore, that socio-legal study be able to drill down to national and sub-national legislative and juridical particulars" (2008, 184). Accordingly, before we can get on to critiquing the RMA's offences, in Part III of this book, it is necessary to explain how the RMA works and then develop a theoretical framework for conducting the assessment. These tasks are undertaken in this chapter and the following one (respectively). Specifically in this chapter, the RMA is first discussed in the context of environmental regulation, before the offences, penalties and institutional arrangements for prosecuting are described. Finally, the RMA is compared and contrasted to international approaches to environmental non-compliance.

M. Wright, *Responding to Environmental Crimes*, Palgrave Studies in Green Criminology, https://doi.org/10.1007/978-3-030-89250-0_2

Regulation and the RMA

A common starting point for considering the term regulation is Selznick's "sustained and focused control exercised by a public agency over activities that are valued by a community" (1985, 363). While, for reasons that will be discussed in the next paragraph, this definition is now considered to be "highly problematic" (Baldwin et al. 2010, 12), it provides some useful indicators as to what is traditionally meant by the term. For instance, sustained and focused control suggests that regulation is something that happens on purpose, rather than occurring unintentionally. Further, the reference to the control being exercised by a public agency presupposes that it is the state that is regulating. Finally, because the activities being regulated are "valued", the goal is not necessarily to stop the behaviour (although this will sometimes be necessary), but rather to modify it to accord with collective goals (Yeung 2004, 6). This qualification also makes it clear that the subject matter is generally considered different to that found in the traditional areas of criminal law and the criminal justice system, where the government is still controlling but it is over non-valued activities (Ogus 1994, 1).

As just stated, in more recent years issues have been identified with Selznick's definition of regulation. Perhaps the greatest concern has been the realisation that it is not just the government that is "regulating" behaviour—third parties have an influence and businesses themselves have a role to play, and indeed can often perform the task more efficiently than government (see generally Gunningham et al. 1998). However, another concern has been that regulation may not be happening intentionally, but rather could be occurring as a side effect of pursuing other objectives (Baldwin et al. 2011, 3). Such issues have led to the word regulation being given a number of meanings in modern usage, with Black noting that the main textbooks contain three types of definition of regulation: a narrow definition of rules created by government and enforced by a public agency, a wider definition retaining government as the regulator but expanding the techniques to any form of direct state intervention in the economy and the widest being "all mechanisms of social control or influence affecting behaviour from whatever source, whether intentional or not" (Black 2001, 129). It is sufficient to note at this point that the RMA is obviously an example of Government intervention (thus ruling out the widest definition).

The reasons a Government might regulate, in the environmental context, can broadly be split into economic and (what have been termed) "ethical" reasons (Hilson 2000, 3–5). The latter includes anthropocentric theories—for instance, that we should regulate "not only in order to protect human health, but also to protect nature because of its instrumental value to humans in providing drugs, food, recreation, spiritual and other beliefs" (Hilson 2000, 3)—as well as biocentric theories, which place humans at the same level as other species, and ecocentric theories, which see humans as just one part of nature as a whole (Brisman and South 2019, 4–5). More common, however, are economic reasons for regulating, with the traditional justification being that there has been market failure which, together with private law failure, suggests a prima facie case for regulatory intervention in the public interest (Ogus 1994, 30).

There are a number of reasons that markets fail, but in the environmental field the "failure" is generally attributed to what are known as externalities (Lin 2006, 908–909). These occur when the production of goods or services imposes costs on third parties that are not reflected in the price charged (Ogus 1994, 35). There may, of course, be private solutions available to affected parties to deal with the effects of externalities, but such solutions bring with them their own issues. For instance, diffuse discharges of contaminants may make it difficult to identify a specific victim and, even where there is an identifiable victim, the high transaction costs involved may mean that it is not practical for every affected person to bring civil proceedings against the party at fault (Bowles et al. 2008). This, therefore, generally leads to regulation to require the internalisation of these costs that are being imposed on others, which is known as the "polluter pays" principle (Baldwin et al. 2011, 18).

The economic, market failure, rationale is adopted by the RMA, the Act being a classic example of regulating due to the presence of externalities (Pickford 2014, 41; Memon 1993, 96). As the Minister for the Environment, the Hon Simon Upton, said on the third reading of the Resource Management Bill:

> In adopting the present formulation of clause 4 the Government has moved to underscore the shift in focus from planning for activities to regulating their effects of which I have spoken. We run a much more liberal market economy these days. Economic and social outcomes are in the hands of citizens to a much greater extent than they have previously been. The Government's focus is now on externalities – the effects of those activities on the receiving environment – and those effects have too often been ignored. ((4 July 1991) 516 NZPD 3019)

The reference to clause 4, in the quote above, is to the purpose of the RMA, and this clause became section 5 when the Bill was enacted. This section sets out that the purpose of the Act is "to promote the sustainable management of natural and physical resources" (subsection (1)), with "sustainable management" defined in subs (2) to mean:

> ... managing the use, development, and protection of natural and physical resources in a way, or at a rate, which enables people and communities to provide for their social, economic, and cultural well-being and for their health and safety while—
>
> (a) sustaining the potential of natural and physical resources (excluding minerals) to meet the reasonably foreseeable needs of future generations; and
> (b) safeguarding the life-supporting capacity of air, water, soil, and ecosystems; and
> (c) avoiding, remedying, or mitigating any adverse effects of activities on the environment.

Two points can be made on this. First, the term "sustainable management" was chosen, instead of the more well-known term "sustainable development" that had been popularised by the Brundtland Report (World Commission on Environment and Development 1987), because the latter term was thought to embrace "a very wide scope of matters including social inequities and global redistribution of wealth", and it was said to be "inappropriate for legislation of this kind to include such goals" (New Zealand Review Group on the Resource Management Bill 1991, 6). Second, the purpose is supported by three principles in sections 6,

7 and 8 of the RMA, which relate to eight matters of national importance, 11 "other matters" (such as the Māori concept of kaitiakitanga, which broadly translates to guardianship and protection, the efficient use and development of natural and physical resources and the effects of climate change) and the principles of the Treaty of Waitangi (New Zealand's founding document between the Crown and 540 Māori chiefs (see generally Ministry for Culture and Heritage 2017)).

How is sustainable management to be achieved? Baldwin et al. note that:

> If the state wants to control, say, the pollution of a river, it may approach the issue in a number of ways. It may decide to regulate the pollution directly by means of a government department or agency; it may rely on the polluting firms to self-regulate (perhaps under state oversight); or it might delegate the control function to third parties such as public interest groups and the commercial partners of the polluters. (2011, 105)

The RMA adopts the first of these approaches—direct state intervention—and does so in two ways: "command and control" and the use of economic instruments. The latter, which have historically been the less common of the two (Keohane et al. 1998), seek to influence behaviour by, for example, rewarding those that meet targets and taxing those that do not. During the deregulation period of the late 1980s and early 1990s there was an increase in their use, as they were considered to be more efficient than command and control regulation (Gunningham 2009, 185), and the RMA provided the legislative framework for their introduction, albeit in a limited way (Buhrs and Bartlett 1993, 107). However, whether due to the limited nature of the economic instruments or for other reasons, there was little uptake of these either in the early years following the RMA's enactment (Richardson 1998, 28), or more recently (Palmer 2015, 3.102), and they are not considered any further in this book.

Command and control regulation involves the use of government-imposed rules ("commands") backed by negative sanctions ("controls"). In terms of "commands", these can be found in legislation, regulations or rules (or a combination of these). There are a number of different ways the "command" can be drafted, but a typical approach is to either create an absolute prohibition or use a standard that must be met. Different types of standards include specification standards (for example, the plant must be designed to certain standards), performance standards (for example, the plant must be run to certain standards) or outcome standards (for example, the emissions from the plant must meet certain standards) (Baldwin et al. 2011, 296–298). Further, the command may be coupled with a prior approval mechanism—for instance, the activity is only prohibited or the standard must only be met if the regulatee has not previously been given permission to breach it by way of a licence (Abbot 2009, 4). In terms of "controls", the traditional approach has been to make the breach of a command a criminal offence (Baldwin 1997, 65), although regulators have long held the power to apply compliance measures and incapacitation mechanisms (and it is typically also a criminal offence if you fail to comply with these), and alternative (that is, non-criminal) sanctions are popular in some countries (see generally Macrory 2006).

The offences this book critiques fit squarely within command and control regulation. Part 3 of the Act sets out the duties and restrictions imposed on regulatees (the "commands"). These relate to the use of land, subdivision, the coastal marine area, the beds of rivers and lakes, the taking and use of water and discharges of contaminants (RMA, sections 9 and 11 to 15). For some of these duties and restrictions, such as the use of land (section 9), the starting point is that the use is permitted unless it contravenes a national environmental standard, a regional rule or a district rule. Further, even if the use is in contravention of a standard or rule (as the case may be), it will still be permitted if expressly allowed by a resource consent or an existing use. For other duties and restrictions, such as the discharge of contaminants, the position is reversed—the activity is prima facie prohibited and only permitted if expressly allowed by a national environmental standard or other regulations, a rule in a regional plan as well as a rule in a proposed regional plan for the same region (if there is one), or a resource consent.

Before moving on to the "controls", it is important to be clear about two things. The first is the specific roles of central government and local authorities (and the different roles within the latter) in this process, these being outlined in Part 4 of the Act. Central government's role, among other things, is to set overall guidance, by way of national policy statements, national environmental standards and national planning standards (RMA, section 24(a)–(ba)), and to monitor the effect and implementation of the RMA, the relationship between the functions, powers and duties of central government and local government under this Part, and any matter of environmental significance (RMA, section 24(f)–(ga)). Implementation is devolved to local authorities, with responsibilities divided between regional councils and territorial authorities (with unitary authorities, defined in section 5(1) of the Local Government Act 2002 as a territorial authority that has the responsibilities, duties and powers of a regional council conferred on it under the provisions of any Act or an Order in Council giving effect to a reorganisation plan, undertaking both). Regional councils are responsible for, among other things, contaminated land, coastal marine areas, the taking and use of water and discharges of contaminants (RMA, section 30). Territorial authorities are responsible for land use, noise, subdivision and the effects of activities on the surface of rivers and lakes (RMA, section 31).

The second thing to be clear on is the manner in which central and local government make the various standards, policies and plans, and that councils undertake the resource consent process. This is set out in Parts 5 and 6 of the Act (respectively). In particular, Part 5 contains what has been described by the Supreme Court in *Environmental Defence Society Inc v New Zealand King Salmon Co Ltd* [2014] NZSC 38, [2014] 1 NZLR 593 as:

> ... a cascade of planning documents, each intended, ultimately, to give effect to s 5, and to Part 2 more generally. These documents form an integral part of the legislative framework of the RMA and give substance to its purpose by identifying objectives, policies, methods and rules with increasing particularity both as to substantive content and locality. (at [30])

Part 5 also contains section 84(1) of the Act, which provides that "[w]hile a policy statement or a plan is operative, the regional council or territorial authority concerned, and every consent authority, shall observe and, to the extent of its authority, enforce the observance of the policy statement or plan". While drafted somewhat obliquely, a point that is returned to in later chapters, according to the Ministry for the Environment this provision "creates a general duty on councils to implement their own plans and policy statements, which requires councils to carry out [compliance, monitoring and enforcement] to monitor compliance and respond to non-compliance" (Ministry for the Environment 2018a, 13).

It will be recalled that, at the beginning of this section, the widest definition of regulation was ruled out. It can now be observed that the RMA is most closely aligned to the narrowest definition of regulation referred to earlier—rules created by (central) government and enforced by a public (central government) agency—although it does move beyond this in two ways. One is that, in terms of the central government rules, what is specifically prohibited (and allowed) can be modified, subject to certain limits, by the local authorities using their district or regional plans (as the case may be) and the granting of resource consents. The other is that those responsible for enforcement (local authorities), while being a public agency (broadly defined), are quite separate from central government. Having said that, it can be observed that steps have been taken to increase central government involvement with the enactment of the Resource Management Amendment Act 2020, which allows, in certain circumstances, the Environmental Protection Authority (a central government agency) to start investigations, help local authorities and even intervene in an existing investigation.

Turning then, finally, to enforcing the duties and restrictions, there are a number of "controls" set out in Part 12 of the Act. These can be split into directive controls, which require a person to do something (or not do something) to ensure compliance; and punitive controls, which penalise an offender for non-compliance (Ministry for the Environment 2018a, 69). The former includes the abatement notice and the enforcement order, and the latter includes the infringement offence regime and criminal offences, and a local authority can choose to go down one or both routes (simultaneously or one after another).

Abatement notices are a lower level directive measure. They are issued by enforcement officers using a prescribed form (Resource Management (Forms, Fees, and Procedure) Regulations 2003, Form 48) and there are, broadly, two types. The first requires a person to cease doing something that is contravening the RMA, any regulations made under the Act, a rule in a plan or a resource consent (RMA, section 322(1)(a)). The second requires a person to do something to ensure compliance by or on behalf of that person with the Act, any regulations, a rule in a plan or a proposed plan, or a resource consent, but can only be issued where it is also necessary to avoid, remedy, or mitigate any actual or likely adverse effect on the environment caused by or on behalf of the person or relating to any land of which the person is the owner or occupier (RMA, section 322(1)(b)).

An abatement notice can be made subject to such conditions as the enforcement officer serving it thinks fit (RMA, section 322(3)), and may be issued if the enforcement officer has reasonable grounds for believing that any of the circumstances justifying its issuance exist (RMA, section 322(4)). The notice must be complied with before the period specified in it expires (RMA, section 323(1)(a)), and the courts have held that they can be of indefinite duration (*Roberts v Northland Regional Council* [2014] NZHC 284). An appeal against a notice must be made to the Environment Court and this will, usually, not automatically operate as a stay (section 325(3)(a)). Rather, a stay must be granted by an Environment Judge (RMA, section 325(3)(b)).

Enforcement orders (and interim enforcement orders) are higher level directive measures. They are made by the Environment Court, following either an application with a supporting affidavit(s) (RMA, section 316 and Resource Management (Forms, Fees and Procedure) Regulations 2003, Form 43) or on sentencing (RMA, section 339(5)(a), as discussed further in the next section). An enforcement order covers not only the same general ground as an abatement notice, but other matters. For instance, it can require a person to remedy or mitigate any adverse effect on the environment caused by or on behalf of that person (RMA, section 314(1)(c)). Or it can require a person to pay money to or reimburse any other person for any actual and reasonable costs and expenses which the other person has incurred or is likely to incur in avoiding, remedying or mitigating any adverse effect on the environment, although this is only where the person against whom the order is sought fails to comply with an order under any other paragraph of section 314(1), an abatement notice, a rule in a plan or a proposed plan or a resource consent or any of that person's other obligations under the Act (section 314(1)(d)). Most enforcement orders can be applied for by "any person", although some, for instance, where it relates to changing or cancelling a resource consent, are restricted to local authorities (RMA, sections 314(1)(e) and 316(2)).

Infringement offences are a lower level punitive response. The infringement offence regime is engaged by an enforcement officer who has observed a person committing an infringement offence, or who has reasonable cause to believe such an offence is being or has been committed by that person, serving that person with an infringement notice (RMA, section 343C(1)). Penalties are fixed and range from $300 to $1,000 (Resource Management (Infringement Offences) Regulations 1999), with no conviction being entered. Since 2020, the maximum infringement fee that can be set by regulations has been split between natural and non-natural persons, it now being $2,000 for the former and $4,000 for the latter (RMA, section 360(1)(bb)). The regulations themselves have not, though, been amended as yet. If the infringement fee is paid, the matter is at an end. However, if it is not paid within 28 days the local authority can issue a reminder notice, and if it remains unpaid after a further 28 days then the local authority can file the notice in Court. A hearing will then be held to determine whether or not the offence has been committed. If so, the Court is not restricted to the infringement fee when imposing the penalty (*Otago Regional Council v Bloem* [2010] NZRMA 322 (DC)).

Infringement offences were not included in the RMA when it was first enacted, but were introduced by the Resource Management Amendment Act 1996. When the amendment bill that brought in the infringement offence regime was introduced, it was put forward by the Hon Simon Upton MP, the Minister for the Environment, as "the only sensible cost-effective way of enabling minor breaches to be dealt with swiftly"; otherwise, it was said, such breaches would not be enforced ((14 December 1995) 552 NZPD 410–411). Almost all of the offences in section 338 of the RMA (as discussed next) were (ultimately) made infringement offences, with breach of an enforcement order being the main exception. In 2012, the Supreme Court decided that the RMA's infringement offence regime was "standalone" (*Down v R* [2012] NZSC 21, [2012] 2 NZLR 585). This was instead of it using the infringement notice process set out in section 21 of the Summary Proceedings Act 1957, with only aspects of this process incorporated by reason of section 343C(4) of the RMA. This is important because it means that, rather than requiring the leave of a District Court Judge or Registrar before a prosecution can be commenced, those enforcing the provisions of the RMA can decide themselves whether to commence a prosecution or issue an infringement notice, such choice being "entirely a matter of prosecutorial judgment in every case" (*Down v R* at [30]).

The final "control" is the offences, the higher level punitive response, and these can be separated into three "categories" (Warnock and de Silva 2015, 1014–1015). First, breaches of duties and restrictions and breaches of enforcement orders, abatement notices (other than for unreasonable noise) and water shortage directions (RMA, sections 338(1), (1A) and (1B)). Second, failures to provide information to enforcement officers and to protect sensitive information, and breaches of excessive noise directions, abatement notices for unreasonable noise and other Environment Court orders (RMA, section 338(2)). Third, obstruction of enforcement processes and breaches of provisions relating to esplanade or access strips (RMA, section 338(3)).

The second and third categories of offence are not considered in this book, for two reasons. One is that they are less serious than the offences in the first category; the maximum penalty for those in the second category is a fine not exceeding $10,000 (RMA, section 339(2)) and the maximum penalty for those in the third category is a fine not exceeding $1,500 (section 339(3)). The other is that they are prosecuted very infrequently. For instance, in the period 1 July 2008 to 30 September 2012 there were only three prosecutions under section 338(3)(a) (for obstructing enforcement officers) and no prosecutions for contraventions of section 338(2) or sections 338(3)(b) or (c) (Ministry for the Environment 2013, 11 and 13). It is not clear from the Ministry for the Environment's National Monitoring System data how many of these prosecutions were there in 2014/2015, 2015/2016 or 2016/2017, as these fall under the heading "other" (which also includes breaches of an abatement notice and an enforcement order), but there were only 14 "other" prosecutions in 2014/2015, 1 "other" prosecution in 2015/2016 and 2 "other" prosecutions in 2016/2017 (Ministry for the Environment 2016, 2017, 2018b). All references to "offences" from this point are solely to those in the first category, and in particular to breaches of duties and restrictions and breaches of enforcement orders and abatement notices.

The RMA's Offences, Penalties and Institutional Arrangements for Prosecuting

In terms of the RMA's offences, the primary provision for the purpose of this book is section 338(1):

> Every person commits an offence against this Act who contravenes, or permits a contravention of, any of the following:
>
> (a) sections 9, 11, 12, 13, 14, and 15 (which impose duties and restrictions in relation to land, subdivision, the coastal marine area, the beds of certain rivers and lakes, water, and discharges of contaminants):
> (b) any enforcement order:
> (c) any abatement notice, other than a notice under section 322(1)(c):
> (d) any water shortage direction under section 329.

Section 338(1) is supported by two other provisions in the Act. One is section 340, which provides for both vicarious liability and, where a non-natural person has been convicted, a mechanism to also convict directors and those involved in the management of the defendant. In terms of vicarious liability, there are defences for the party that is prima facie liable if it can prove two things. First, it must prove either that it did not know, and could not reasonably be expected to have known, that the offence was to be or was being committed, or it took all reasonable steps to prevent the commission of the offence. Second, the defendant also needs to prove that it took all reasonable steps to remedy any effects of the act or omission giving rise to the offence. In terms of the mechanism to convict directors and those involved in the management of the defendant, the prosecution must prove two things. First, that the act or omission that constituted the offence took place with his or her authority, permission or consent. Second, that he or she knew, or could reasonably be expected to have known, that the offence was to be or was being committed and that he or she failed to take all reasonable steps to prevent or stop it.

The other provision that supports section 338(1) is section 341, which is titled "Strict liability and defences". It provides, in subsection (1), that in a prosecution for an offence of contravening or permitting a contravention of any of sections 9, 11, 12, 13, 14 and 15, the prosecution does not have to prove that the defendant intended to commit the offence. This is subject to two defences, one relating to necessity and the other to events beyond the control of the defendant (subsection (2)). In both cases, the effects of the action or event must have been adequately mitigated or remedied by the defendant after it occurred. If these defences are to be relied on, they must be notified to the prosecution, together with the facts that support the defence, within 7 days after the service of the summons (or longer if the Court allows) (subsection (3)).

In terms of penalties, the main provision is section 339(1). When the RMA was enacted, the maximum penalty was imprisonment for a term not exceeding 2 years or a fine not exceeding $200,000. However, in 2009 the penalties were split between natural and non-natural persons, and increased (Resource Management (Simplifying and Streamlining) Amendment Act 2009), such that the provision now reads:

> Every person who commits an offence against section 338(1), (1A), or (1B) is liable on conviction,—
>
> (a) in the case of a natural person, to imprisonment for a term not exceeding 2 years or a fine not exceeding $300,000:
> (b) in the case of a person other than a natural person, to a fine not exceeding $600,000.

These penalties are then augmented by the rest of section 339 and by section 342. In terms of the former, where the offence is a continuing one the maximum penalty increases by $10,000 for "every day or part of a day during which the offence continues" (RMA, section 339(1A)).

Section 339 also provides, in subsection (4), that the Court can sentence natural persons to community work, although strictly speaking this is not necessary as the Sentencing Act 2002 (which applies to all criminal offending in New Zealand) already allows the Court to impose this sentence (along with various other sentences). Finally, the Court may, instead of or in addition to imposing a fine or a term of imprisonment, make an enforcement order or, if the offence involves an act or omission that contravenes a resource consent, make an order requiring a consent authority to serve notice of the review of a resource consent held by the person (subsection (5)). In terms of section 342, this provides, as a starting point, that any fine imposed is to be paid to the local authority that commenced the proceeding (subsection (1)). However, the Crown will retain 10% of the fine (subsection (2)), except "where any money awarded by a court in respect of any loss or damage is recovered as a fine, and that fine is ordered to be paid to a local authority under subsection (1), no deduction shall be made under subsection (2) in respect of that money" (subsection (3)).

In terms of the institutional arrangements for prosecutions, the implication from section 342 is that charges are to be brought by local authorities. However, this provision only outlines the consequences of what happens when a council brings a (successful) case, rather than a requirement that it does so. Instead, as noted in the section above, the main direction to local authorities regarding enforcement is in section 84(1) of the Act, which provides that "[w]hile a policy statement or a plan is operative, the regional council or territorial authority concerned, and every consent authority, shall observe and, to the extent of its authority, enforce the observance of the policy statement or plan".

Finally, section 309 provides that RMA prosecutions are heard in the District Court but, unless otherwise directed by the Chief District Court Judge, by an Environment Judge (as opposed to a generally warranted District Court Judge). Environment Judges are District Court Judges and are appointed by the Governor-General on the recommendation of the Attorney-General, after consultation with the Minister for the Environment and the Minister of Māori Affairs. Provision is also made for the appointment of alternate Environment Judges, who may act as an Environment Judge when the Principal Environment Judge, in consultation with the Chief District Court Judge or Chief Māori Land Court Judge, considers it necessary for the alternate Environment Judge to do so.

In summary, the RMA regulates in a (primarily) command and control manner, with the commands being coupled with a prior approval mechanism and the controls ranging from low-level directive measures (issued by enforcement officers) to high-level Court-based punitive responses. The latter are criminal offences that include vicarious liability, deeming provisions to ensnare directors and management and liability without fault, and penalties include fines and imprisonment (and everything in between), as well as enforcement orders and resource consent reviews. Finally, local authorities are considered responsible for bringing prosecutions and receive the overwhelming majority of any fine imposed, with such prosecutions heard by specialists who sit in general courts. Before moving on to set out the theoretical aspect to the framework, in Chapter 3, it is worth considering how the foregoing compares to international approaches.

International Comparisons

Much of what has just been outlined is unremarkable internationally. Many countries regulate the environment in a command and control manner, with enforcers having a range of controls at their disposal, including the criminal law. For instance, Abbot's review of New South Wales and Victoria (Australia), Ontario (Canada) and England and Wales found that:

> Despite the institutional variations, in all three countries, 'command and control' regulation is at the core of the legislative framework. Prior approval (secured by the issuing of a licence or permit) is required for certain regulated activities and operations are controlled through the use of licence conditions incorporating inter alia environmental standards and operating requirements. Non-compliance with the regulatory controls is underpinned by criminal liability and regulators can utilise a variety of compliance and enforcement mechanisms in responding to violations. (2009, 88–89)

In terms of the criminal law specifically, it is also very common to find some combination of vicarious liability, deeming provisions to ensnare directors and management and liability without fault, although some jurisdictions (such as various states in Australia) tier their offences, with greater sanctions for *mens rea* offences (see generally Abbot 2009, 119–152).

The RMA's approach to penalties is also common internationally, with environmental statutes typically allowing the courts to impose both fines and imprisonment (Billiet and Rousseau 2014), although the availability of additional sentences is more mixed. Some jurisdictions, such as New South Wales (Australia), contain extensive sentencing options, including orders that the defendant publish details of the offending or that the defendant carry out an environmental audit of activities it undertakes (Preston 2007, 159–163; White 2018). On the other hand, such orders have not traditionally been available to the courts in England and Wales (Bell et al. 2017, 292–294).

When it comes to the institutional arrangements, matters start to diverge from the New Zealand approach. First, in terms of compliance, monitoring and enforcement, it is common internationally to find enforcement remaining more centralised, often through an Environmental Protection Authority (see generally White and Heckenberg 2014, 218–222). A corollary of this is that such agencies often focus exclusively (or almost exclusively) on compliance, monitoring and enforcement, compared to the New Zealand approach where this is only one task out of many.

Second, in terms of judges, specialists sitting in specialist courts are by no means unheard of. For instance, Pring and Pring found that, as of 2012, there was approximately 465 specialised environmental courts and tribunals in 46 countries (Pring and Pring 2011, 484). Perhaps the best-known example, in the criminal context, is the New South Wales Land and Environment Court, which has been described as a "*problem-solving court*" concerned with repairing "*environmental harm* where possible and feasible" (White 2017, 122, emphasis in original). Indeed, while cautioning that "[c]are should be taken to not romanticise or over-inflate" its positive work, White has suggested that this Court is "at the cutting edge of 'what could be' and 'what should be'" (2017, 130).

However, it is unusual to see specialist judges sitting in generalist courts, as Environment Judges do when adjudicating on RMA prosecutions (in other matters they sit in the Environment Court, a standalone national court that sits in a number of different parts of the country). What this means is that, unless legislation specifically provides for the manner in which a District Court is to deal with an RMA prosecution (which it does to only a very limited extent), the usual rules of criminal procedure apply. For instance, a prosecution under the RMA is commenced by filing a charging document using the procedure set out in the Criminal Procedure Act 2011, the prosecutor must then comply with the requirements of the Criminal Disclosure Act 2008 and so on.

The other area where we see an even greater divergence from international environmental regulation statutes is the RMA's provision requiring that (90% of) the fines imposed are returned to the prosecuting local authority. Generally, fines imposed in environmental prosecutions in other jurisdictions go into the Crown's consolidated fund, or at best are directed towards an environmental fund or a specific environmental project (see, in the context of New South Wales (Australia): White 2017, 128; Westerhuis 2013, 211). The objection to fines being returned to the prosecutor is not often stated. However, in Macrory's examination of the system of regulatory sanctions in the United Kingdom he was of the view, when discussing how regulators should "Avoid perverse incentives that might influence the choice of sanctioning response", that "regulators should not retain the revenue from Monetary Administrative Penalties, or exercise any control over how that revenue should be spent" (2006, 33). Bishop, although not being against fines being returned to the regulator, offers another reason to be wary:

> It is possible that such a development could lead to the misuse of prosecutorial discretion. An enforcement agency, cognisant of funding constraints, may well target those offenders who have the financial resources easily to pay fines and/or administrative monetary penalties; solvent offenders are not necessarily the most serious polluters and,

as such, the hypothecation of penalty receipts may operate counter-productively, with the environment suffering as a result. (2009, 302).

If the concern is meeting the costs of the prosecutor, a more common approach, found in various Australian states, is to allow the sentencing Judge to make a separate award of costs (see, in the context of New South Wales (Australia): White 2017, 128).

References

Abbot, Carolyn. 2009. *Enforcing Pollution Control Regulation*. Portland: Hart Publishing.

Baldwin, Robert. 1997. Regulation: After 'Command and Control'. In *The Human Face of Law: Essays in Honour of Donald Harris*, ed. Keith Hawkins. Oxford: Clarendon Press.

Baldwin, Robert, Martin Cave, and Martin Lodge. 2010. Introduction: Regulation—The Field and the Developing Agenda. In *The Oxford Handbook of Regulation*, ed. Robert Baldwin, Martin Cave, and Martin Lodge, 3–14. New York: Oxford University Press.

Baldwin, Robert, Martin Cave, and Martin Lodge. 2011. *Understanding Regulation*, 2nd ed. New York: Oxford University Press.

Bell, Stuart, Donald McGillivray, Ole W. Pedersen, Emma Lees, and Elen Stokes. 2017. *Environmental Law*, 9th ed. Oxford: Oxford University Press.

Billiet, Carole M., and Sandra Rousseau. 2014. How Real Is the Threat of Imprisonment for Environmental Crime? *European Journal of Law and Economics* 37 (2): 183–198. https://doi.org/10.1007/s10657-011-9267-2.

Bishop, Patrick. 2009. Criminal Law as a Preventative Tool of Environmental Regulation: Compliance Versus Deterrence. *Northern Ireland Legal Quarterly* 60 (3): 279–304.

Black, Julia. 2001. Decentring Regulation: Understanding the Role of Regulation and Self-Regulation in a 'Post-Regulatory' World. *Current Legal Problems* 54 (1): 103–146.

Bowles, Roger, Michael Faure, and Nuno Garoupa. 2008. The Scope of Criminal Law and Criminal Sanctions: An Economic View and Policy Implications. *Journal of Law and Society* 35 (3): 389–416. https://doi.org/10.1111/j.1467-6478.2008.00444.x.

Brisman, Avi, and Nigel South. 2019. Green Criminology and Environmental Crimes and Harms. *Sociology Compass* 13 (1). https://doi.org/10.1111/soc4.12650.

Buhrs, Ton, and Robert V. Bartlett. 1993. *Environmental Policy in New Zealand: The Politics of Clean and Green?* Auckland: Oxford University Press.

Gunningham, Neil. 2009. Environment Law, Regulation and Governance: Shifting Architectures. *Journal of Environmental Law* 21 (2): 179–212.

Gunningham, Neil, Peter Grabosky, and Darren Sinclair. 1998. *Smart Regulation: Designing Environmental Policy*. Oxford: Clarendon Press.

Hilson, Chris. 2000. *Regulating Pollution: A UK and EC Perspective*. Oxford: Hart Publishing.

Keohane, Nathaniel, Richard Revesz, and Robert Stavins. 1998. The Choice of Regulatory Instruments in Environmental Policy. *Harvard Environmental Law Review* 22: 313–367.

Lin, Albert. 2006. The Unifying Role of Harm in Environmental Law. *Wisconsin Law Review* 2006 (3): 897–985.

Macrory, Richard. 2006. *Regulatory Justice: Making Sanctions Effective*.

Memon, P. Ali. 1993. *Keeping New Zealand Green: Recent Environmental Reforms*. Dunedin: University of Otago Press.

Ministry for Culture and Heritage. 2017. The Treaty in Brief. https://nzhistory.govt.nz/politics/treaty/the-treaty-in-brief. Accessed 22 April 2021.

Ministry for the Environment. 2013. A Study into the Use of Prosecutions Under the Resource Management Act 1991: 1 July 2008–30 September 2012.

Ministry for the Environment. 2016. National Monitoring System for 2014/15.

Ministry for the Environment. 2017. National Monitoring System for 2015/16.

Ministry for the Environment. 2018a. Best Practice Guidelines for Compliance, Monitoring and Enforcement under the Resource Management Act 1991.

Ministry for the Environment. 2018b. National Monitoring System for 2016/17.

New Zealand Review Group on the Resource Management Bill. 1991. Report of the Review Group on the Resource Management Bill.

Ogus, Anthony. 1994. *Regulation: Legal form and Economic Theory*. Oxford: Clarendon Press.

Palmer, Kenneth. 2015. Resource Management Act 1991. In *Environmental and Resource Management Law*, ed. Derek Nolan. Wellington: LexisNexis.

Pickford, Michael. 2014. The 'Environmental Bottom Line' and Economic Efficiency. *Canterbury Law Review* 20: 39–51.

Preston, Brian. 2007. Principled Sentencing for Environmental Offences—Part 2: Sentencing Considerations and Options. *Criminal Law Journal* 31: 142–165.

Pring, George, and Catherine Pring. 2011. The Future of Environmental Dispute Resolution. *Denver Journal of International Law and Policy* 40 (1–3): 482–491.

Richardson, Benjamin. 1998. Economic Instruments and Sustainable Management in New Zealand. *Journal of Environmental Law* 10 (1): 21–39.

Selznick, Philip. 1985. Focusing Organisational Research on Regulation. In *Regulatory Policy and the Social Sciences*, ed. Roger Noll, 363–364. Berkley: University of California Press.

Warnock, Ceri, and Karenza de Silva. 2015. Compliance and Enforcement. In *Environmental law in New Zealand*, ed. Peter Salmon and David Grinlinton, 977–1038. Wellington: Thomson Reuters.

Westerhuis, Diane. 2013. A Harm Analysis of Environmental Crime. In *Emerging Issues in Green Criminology: Exploring Power, Justice and Harm*, ed. Reece Walters, Diane Westerhuis, and Tanya Wyatt, 197–217. Basingstoke: Palgrave Macmillan.

White, Rob. 2008. *Crimes Against Nature: Environmental Criminology and Ecological Justice*. Devon: Willan Publishing.

White, Rob. 2017. Reparative Justice, Environmental Crime and Penalties for the Powerful. *Crime, Law and Social Change* 67 (2): 117–132. https://doi.org/10.1007/s10611-016-9635-5.

White, Rob. 2018. Sentencing Environmental Offenders: It Is Not Just About the Money. In *Green Crimes and Dirty Money*, ed. Toine Spapens, Rob White, Daan van Uhm, and Wim Huisman. Abingdon: Routledge.

White, Rob, and Diane Heckenberg. 2014. *Green Criminology: An Introduction to the Study of Environmental Harm*. Abingdon: Routledge.

World Commission on Environment and Development. 1987. *Our Common Future*. Oxford: Oxford University Press.

Yeung, Karen. 2004. *Securing Compliance: A Principled Approach*. Oxford: Hart Publishing.

3

Regulatory Enforcement and the Criminal Law

It can be observed, from Chapter 2, that the offences that are the subject of this book engage two overlapping theoretical areas. First, they are a regulatory enforcement mechanism, being one way in which local authorities can respond to breaches of the RMA. Second, they rely on the criminal law for determining whether a breach has occurred and what penalty, if any, should be imposed. These two areas—regulatory enforcement and the criminal law—are, accordingly, the subject of this chapter, with a view to developing a framework against which the use of the RMA's offences can be assessed.

Regulatory Enforcement Theory

Let us start with regulatory enforcement. Or, more accurately, let us start with regulatory compliance, as it is only once one understands the reasons why some regulatees comply with rules, and others do not, that any progress can be made on enforcement. This is, therefore, the

M. Wright, *Responding to Environmental Crimes*, Palgrave Studies in Green Criminology, https://doi.org/10.1007/978-3-030-89250-0_3

first task in this section, before consideration turns to how a regulator should detect and respond to non-compliance, and then why, according to regulatory scholars, sanctions should be imposed.

Reasons for Compliance and Non-compliance

There are a number of reasons why people and businesses comply, or do not comply, with regulatory rules. Perhaps the most straight-forward reason for compliance is where it is in the regulatee's interests to do so (Hutter 1997, 182). This is best seen in areas where a breach of a regulatory standard may have catastrophic consequences that put the regulatee out of business, such as oil refineries and chemical works (Genn 1993, 223). A regulatee may also comply out of self-interest when a rule is worded such that, for the firm in question, it is the most efficient use of resources (Abbot 2009, 31). However, it might be expected that this would happen infrequently, as regulatory rules are often drafted at a high level of generality.

Another group of regulatees choose whether or not to comply with rules based on their internal values (Burby and Paterson 1993, 756; Situ and Emmons 2000, 123). Their obedience is determined by their general moral principles—a civic sense of duty to obey laws—and a specific evaluation of a particular rule, which is made up of the rule's reasonableness, how it was enacted and the approach of the authorities in enforcing it (Winter and May 2001, 677–678). These factors will operate at different strengths; some regulatees' general commitment being so strong that they will follow all but the most outrageous rules, while compliance by other regulatees will be heavily dependent on the appropriateness of a particular rule and how it was made.

The above reasons for compliance broadly fall into what could be called "affirmative motivations", or "good intentions and a sense of obligation to comply" (May 2004, 42). On the other hand, there are reasons that could be called "negative motivations", which are "fears of the consequences of being found in violation of regulatory requirements" (May 2004, 42). The regulatees that comply for these latter reasons are often called "rational utility maximisers" (Faure and Svatikova 2012, 257) or, more pejoratively, "amoral calculators" (Kagan and Scholz 1984, 67–68). This is because they do not consider the rightfulness or wrongfulness of complying (either generally or in a specific case) and instead assess their expected gain from violating a regulatory law against the expected cost of doing so, complying only if the latter is higher than the former. The expected gain may, for instance, be the money saved by not investing in equipment that will prevent a contravention, while the expected cost includes formal and informal sanctions imposed by, or following a process commenced by, the regulator.

Finally, it is necessary to mention two further groups, for whom compliance (or lack thereof) is not necessarily an explicit choice. The first is those regulatees that, despite being willing to comply, simply cannot due to technological, economic or practical limitations. For example, it may be that the pollution abatement machinery required to comply with the regulatory rules is too expensive for the small owner-operated company. Or a medium-sized business is so hopelessly disorganised that, despite having the funds available, it cannot manage the task of acquiring and installing necessary safety equipment. Such regulatees are often considered "unfortunate", rather than wrongful, which can affect how they are dealt with by enforcement officers (Hawkins 1984, 112). Second, some regulatees fail to comply because they do not know the rules. Of these, a portion will be well-intentioned, while others will simply not care, and the assessment of which by the enforcement officer will likely elicit a different response (Baldwin 1990, 324). In particular, enforcement officers may be inclined to deal with the former using informal, low-level enforcement responses (such as education and assistance), whereas the latter are more likely to face formal sanctions.

Dealing with Non-compliance

Turning then to address enforcement squarely, it is often the case that, when new regulation is designed, enforcement is given low priority by government policy analysts when compared to the preparation of the substantive rules, with the tendency being to assume that compliance will be "perfect" (McKean 1980, 270). However, as will be apparent, this assumption is generally not well-founded. In fact, given there are many reasons not to comply with regulation, enforcement is of critical importance—as Gunningham notes, "[e]ffective enforcement is vital to the successful implementation of social legislation, and legislation that is not enforced rarely fulfils its social objectives" (2010, 120).

A critical problem with enforcement, however, is that it is often expensive and budgets are typically limited. Enforcement officers must be employed to detect and respond to breaches; scientific and legal analysis may be required to confirm whether a breach has occurred, and there are costs associated with responding to breaches. Further, as a general rule, this response cost will increase with the level of formality—issuing a compliance notice will be more expensive than a warning, prosecuting an offender will cost more than issuing a compliance notice and so on. This cost is compounded by restricted budgets, with enforcement often occurring in the context of being only one part of a regulator's role and many branches of a public agency also seeking access to scarce resources. This means that regulators have to make decisions as to where to intervene and how to intervene. However, before moving on to these questions, it is necessary to ask why then, in light of these factors, enforcement is nevertheless typically the role of a public agency, rather than undertaken privately.

Public enforcement of regulation should not be confused with the question of why there is public regulation (for instance, rules backed by negative sanctions) instead of relying on private law (for instance, tort claims), which was discussed in Chapter 2. Rather, the question is why a public agency is entrusted with the responsibility of enforcing (public) regulation, instead of enforcement being undertaken privately (for example, by the victim of a regulatory breach). Private enforcement of regulatory laws was the approach historically taken, enforcers being paid "bounties" for the apprehension and conviction of offenders and sharing any fines awarded by the courts with the Crown (Landes and Posner 1975, 2). Even as public agencies began to develop and were given regulatory enforcement powers, the vast majority of prosecutions were still undertaken privately (Ogus 1992, 15–18). This did not change until the middle of the nineteenth century, after which time public agencies increasingly became involved, to the point where they are now almost exclusively responsible for regulatory enforcement (Garoupa et al. 2011, 237–238). In modern times, the public's role is generally limited to reporting suspected violations, giving evidence in court proceedings and, sometimes, pressuring public agencies into taking action.

The shift to public enforcement of regulation can be partially explained by the fact that regulatory breaches often have no identifiable victim. However, that only explains why *victim* enforcement is not required, and not why private parties (for instance, interest groups) cannot perform the task of enforcement (Polinsky and Shavell 2007, 406). Rather, the best explanation is that private enforcement runs the risk of interfering with ongoing relationships between regulators and regulatees, it has the potential for inconsistency and unpredictability, and it may be done incompetently (Gunningham et al. 1998, 105–106). As such, the ideal role is said to be that "citizens would complement, not replace, enforcement by government agencies" (Gunningham et al. 1998, 106).

In the usual course, therefore, enforcement will primarily be undertaken by a public agency, and the first step in securing compliance is for the agency to identify those regulatees that are in breach. This can be done in various ways. A large proportion of an enforcement officer's role is taken up by external, or reactive, tasks (Gunningham and Johnstone 1999, 105), such as responding to breaches reported by a member of the public or declared by the regulated party itself. The latter is not as uncommon as it may seem, as self-reporting of breaches is often a condition of a licence issued by a regulator, and breaching this condition may place the licence in jeopardy. Even if it is not a condition, self-reporting may result in more lenient action by the regulator (for a study on which, see Stretesky 2006). Other ways of identifying those in breach are proactive or internal to the regulator. For instance, it could be found during a random inspection or picked up as part of routine monitoring (for example, of licence conditions).

The dominant view in the regulatory literature is that, in light of scarce resources, it is best to use a risk-based approach to determining the question of where to intervene (Gunningham 2011, 170). Risk-based regulation is concerned with regulatory or institutional risk: "risks to the agency itself that it will not achieve its objectives" (Organisation for Economic Co-operation and Development 2010, 187), which means that deciding where to intervene should be based on the risk of non-compliance and an assessment of the impact such non-compliance will have on the regulator's goals (Black and Baldwin 2010, 181). However, risk-based regulation is not without its issues (see generally Black and Baldwin 2012) and, importantly for present purposes, it does not necessarily assist with the question of how to intervene when a regulator identifies non-compliance (what is often called the regulator's "enforcement strategy" (Windholz 2017, 240)).

For many years, it was believed that regulators were faced with a stark choice when deciding how to secure conformity with the law. Should they primarily take "action to prevent potential law violations without the necessity to detect, process, and penalize violators", or should they mainly be detecting "violations of law, determining who is responsible for their violation, and penalizing violators to deter violations in the future, either by those who are punished or by those who might do so were violators not penalized" (Reiss Jnr 1984, 23)? The two approaches are often called compliance and deterrence (respectively), although such labels can be misleading given both have the same end goal of compliance, differing instead as to the means of getting there (condoning breaches versus penalising them) (Wells 1993, 28).

The compliance approach is favoured by those who see actors as generally good citizens (Abbot 2009, 42). Compliance-minded regulators seek to build relationships with regulatees, the idea being that a co-operative approach will lead to better long-term results. Sanctions are still an option for the regulator, but they operate in the background, as a threat, to be used as a last resort (Diver 1980, 297; Hawkins 2002). Advocates of a compliance approach argue that it is less expensive than relying on formal legal measures, meaning that scarce resources do not get tied up in costly litigation. It also reflects the nature of regulation and its connotations of modifying socially beneficial behaviour (Selznick 1985, as discussed in Chapter 2), as opposed to the outright prohibitions contained in traditional criminal law and dealt with by the police. Opponents, however, claim that a compliance approach is indicative of regulatory capture—the regulatees, rather than the regulator, being in control of the regulation—and that it leads to worse results than doing nothing, as businesses that were otherwise inclined to follow rules would stop doing so once they notice that others are "getting away with it" (Gunningham 2010, 125).

In contrast to a compliance approach, a deterrence strategy takes the position that regulatees are generally rational utility maximisers who need to be inspected regularly and dealt with firmly every time they break the law. This is based, at least in part, on the position that regulatees are usually businesses and thus well versed in weighing up the costs and benefits of taking one course of action over another. Deterrence strategies have close links to the law and economics approach, requiring sanctions to be imposed frequently and at a sufficiently high level to convince regulatees that offending is not worth it. Those in favour of a deterrence strategy note that it reduces the prospect of regulatory capture, as discretion is largely removed, and that it brings consistency and accountability. However, it can be expensive, as resources need to be devoted to inspection, monitoring and formal enforcement action, and can lead to evasion and creative compliance, placing the regulator's ultimate policy goals at risk. Finally, the overall evidence to support such an approach is mixed, with the best that can perhaps be said being that "those who are differently motivated are likely to respond very differently to a deterrence strategy" (Gunningham 2010, 124). However, knowing what drives each regulatee to comply (or not comply) can be problematic (Kagan and Scholz 1984, 86).

These last points set the scene for a broader approach than simply choosing compliance or deterrence, with the literature shifting, towards the end of the twentieth century, from considering regulation and enforcement largely from the perspective of the regulator to considering what motivates the regulated community (Hawkins 2013, 952–954). As noted earlier in this chapter, there are a number of reasons why regulatees comply, and do not comply, with rules, and the enforcement response should therefore take this into account. For instance, a compliance approach can be taken advantage of by rational utility maximisers, while a deterrence approach may provoke the wrong reaction from those "political citizens"that would ordinarily be inclined to follow rules, but deem the one in question to be unreasonable (Kagan and Scholz 1984).

Neither approach may work with the firm that is willing, but unable, to comply—advice will not be of assistance and fining them (and so reducing their available resources) will make it even harder to make the required changes—or with the ignorant, who must instead be educated and informed.

This led to a movement towards more flexible regulatory strategies (of which enforcement is more explicitly only one part), with the seminal work of Ayres and Braithwaite and their concept of "responsive regulation" (1992). Using game theory in the area of regulation, Scholz had previously argued that the best approach for both the regulator and the regulatee is for them to cooperate with one another, and that they should maintain this position until one of them defects (1984). At this point the other party should do the same until the defector returns to cooperation. Building on this, Ayres and Braithwaite theorised that the key to successful cooperative regulation required such a "tit-for-tat" strategy, a hierarchical range of sanctions and interventionism in regulatory style, and a sanctions pyramid with a severely punitive peak (1992, 40). The pyramid, which has been described as the "most distinctive part of responsive regulation" (Braithwaite 2002, 73), may include persuasion and warning letters at the bottom, non-criminal penalties and criminal penalties in the middle, and licence suspension and revocation at the peak (see Fig. 3.1) (Ayres and Braithwaite 1992, 35).

Regulators should start by selecting an enforcement response from the bottom of the pyramid (say, information and education), but, in the event that the regulatee fails to comply, move up to increasingly punitive actions (for example, to a warning letter) (Ayres and Braithwaite 1992, 35–36). If this returns the regulatee to compliance, the regulator should move back down the pyramid; if not, the regulator should continue to escalate (for example, to a non-criminal penalty). Such a process will continue until the regulatee complies or is forced to stop operating by the most punitive response, licence revocation. It can be observed that starting at the bottom of the pyramid provides one way of addressing the problem noted previously, that it is one thing knowing that members of the regulated industry have different motivations, but it is another predicting what drives each firm.

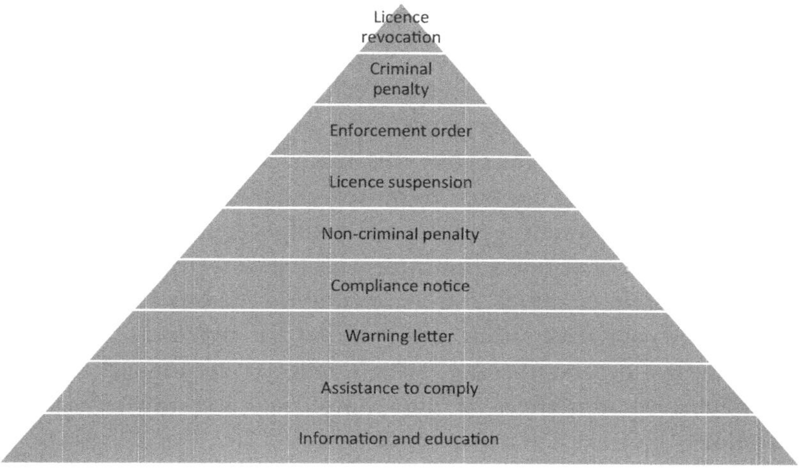

Fig. 3.1 Example enforcement pyramid

Responsive regulation has been very influential in shaping the approach taken by regulators internationally (see, for example, Organisation for Economic Co-operation and Development 2014, 33–36), but it has been criticised on (what have been called) policy/conceptual, practical and principled grounds (see Baldwin and Black 2008, 62–64). Importantly for our purposes, it has also been suggested that responsive regulation splits the criminal justice system in two, with different approaches being taken to offending depending on whether it is the police or a regulatory agency enforcing the statute, rather than the seriousness of the offence (Ashworth 2000, 251). Finally, it has been noted that the focus on ensuring future compliance may not provide reparation for victims harmed by the current offending (Nurse 2016, 86–87).

It could also be said that the enforcement pyramid does not distinguish between penalties, compliance mechanisms and incapacitation measures, with all three being referred to as "sanctions" (see, by way of example, Ayres and Braithwaite 1992, 40). However, the word sanction is normally used in the sense of a penalty, imposed in response to a breach of a (legal) rule (compare Cooter 1984, 1524). By contrast, compliance mechanisms typically require a non-complying regulatee to do, or not do, something, and while they may have a penalising effect their purpose is not to penalise, but to bring the regulated party back into conformity with the law. Incapacitation measures will force a non-complying regulatee to stop operating (either temporarily or permanently), which again may have a penalising effect, but this is not their purpose, with such measures instead being imposed to protect the public.

When the enforcement responses are distinguished in this way, it may be better to think of not one, but three, enforcement pyramids—compliance mechanisms, sanctions and incapacitation measures (see Fig. 3.2). The regulator can select from any pyramid (or pyramids), depending on its reasons for intervening, which reflects, among other things, the different circumstances of non-compliance and the transgressor's motivations.

Responsive regulation has evolved in a number of different directions, including "smart regulation" (Gunningham et al. 1998), a "'two-track' model" (Gunningham and Johnstone 1999), "restorative justice" responsive regulation (Braithwaite 2002) and "really responsive" regulation (Baldwin and Black 2008). It is unnecessary to go into any detail about these strategies, as none have garnered universal acceptance, but the point has been reached whereby responsive regulation and its descendants have become the "regulatory orthodoxy" (Almond and Colover 2012, 1000). Importantly for our purposes, what the approaches have in common is that they are all broadly compliance-based, in that they promote working with the regulatee rather than against them, and they favour having a

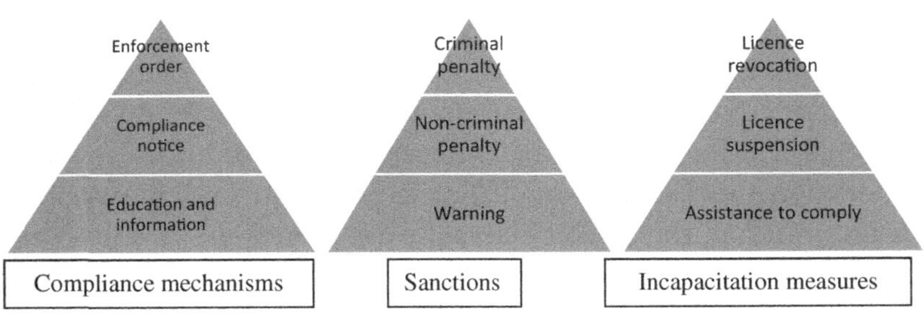

Fig. 3.2 Three enforcement pyramids

number of different enforcement tools and using sanctions only when required.

Primary Rationale for Imposing Sanctions: Deterrence

There are a number of different reasons why a sanction might be imposed. These reasons include because it is deserved (what is known as desert theory), to act as a deterrent (both to the individual being penalised and to the wider regulated community), to provide reparation to those that have suffered losses, to rehabilitate the offender and to incapacitate. However, as Yeung has noted, reparation, rehabilitation and incapacitation are generally not considered appropriate in regulatory matters (2004, 63), and this is even more the case where compliance mechanisms (which are broadly rehabilitative but may also include a reparative element) and incapacitation measures already exist in the regime (but see White 2017).

This leaves desert and deterrence, and regulatory scholars typically suggest that the latter should be the primary rationale for imposing sanctions. For instance, Yeung has observed that:

> Because regulatory law's central concern is to modify behaviour by reference to its perceived undesirable effects on collective welfare, rather than to express moral condemnation of the offence, the theoretical justification for punishment of those who violate regulatory law appears to rest firmly on the deterrence theory of punishment. (1999, 461)

Deterrence—"to prevent or discourage from acting, as by means of fear or doubt" (Silberman 2000, 10524)—has a long history, with it being seen in the eighteenth century writings of Cesare Beccaria and Jeremy Bentham (Beccaria 1764; Bentham 1789). Along with other consequentialist theories, deterrence shares "the idea that punishment is justified by reference to its crime-prevention consequences and [is] advanced within a utilitarian framework" (Roberts and Ashworth 2009, 39). The traditional form of deterrence theory has three components—certainty, severity and celerity (swiftness)—with the idea that:

> Penalties will deter offenders (or potential offenders), so the theory posits, to the extent that they are relatively certain to be imposed, sufficiently severe as to prove aversive and are imposed sufficiently soon after the offence occurs. (Roberts and Ashworth 2009, 40)

As can be seen by the reference to offenders and potential offenders, there are two types of deterrence—specific (individual) and general. The aim of the former is to deter the particular offender that is being punished from offending again, while the aim of the latter is to deter other, would-be, offenders. It is the latter that is generally considered to be the goal of a regulatory sanctioning regime, and what the remainder of this section focusses on.

There are a number of principled criticisms of deterrence—for instance, it would seem to have no issue with punishing an innocent person if it would deter many others, and it treats people as a means to an end—and it fell out of favour for a time. However, it became popular again in the middle of the twentieth century, at least with respect to offences in regulatory statutes, and it is often said that this began with Becker's article *Crime and Punishment: An Economic Approach* (1968). Becker's model starts from the position that people are rational beings who have all the relevant information and weigh up the benefits and costs of acting (1968, 176). Therefore, if sanctions are to deter, their expected cost—the quantum of the penalty ("severity") and the probability of having a penalty imposed ("certainty")—must be greater than the expected benefit from offending.

How to assess the severity of an offence and the certainty it will be imposed will be discussed in Chapter 4, but for now three additional points should be noted. First, the contribution to the expected costs calculation that is provided by the severity of the penalty and the certainty of imposition are not necessarily equal. In particular, the certainty variables are often considered to be more important that the severity of the penalty (Becker 1968; Brown 1992; Gray and Scholz 1993). This must be factored in when determining what the expected cost of an offence will be, although it is impossible to be exact (for instance, it cannot be said that certainty is twice, or three times, or so on, as important as severity).

Second, the requirement that actors be rational can be problematic. This is because, as noted earlier, a number of regulatees will be unaware of the rules and there are a number of reasons why people comply and do not comply. However, there are two (partial) responses to this. One is that regulatees are normally businesses, and therefore likely to have more information than individuals and a greater focus on weighing up costs and benefits (Braithwaite and Geis 1982, 302; Sutherland 1983). The other is that actors are not required to possess "perfect rationality" for the theory to hold, with minimal or limited rationality being enough (Paternoster 2009). Put another way, regulatees need only be "sufficiently rational to be deterrable" (Posner 1985, 1205), which is an easier standard to satisfy.

Third, the reference to "expected" costs (and benefits) means that it is the perceived costs (and benefits), rather than the actual costs (and benefits), that are relevant (Ogus and Abbot 2002, 290). This suggests that it is important that regulatees are aware of actual penalties, how often they are imposed and changes in these when they occur (Paternoster 2010, 804). The last of these is particularly critical if a regulator has introduced a new policy, such as more inspectors to increase the certainty of enforcement action, for without awareness of the changes there can be no additional deterrence (Von Hirsch et al. 1999, 7). Awareness could happen in a number of ways, including as a result of media reports, a regulator's own publicity of enforcement action, or informally, such as during an inspection by an enforcement officer.

Despite certainty generally being considered the more important of the two factors, deterrence theory can imply that the best approach is to impose very severe penalties sporadically (as opposed to minor penalties frequently) (Polinsky and Shavell 2000, 53; Garoupa 1997, 269). This is because, as noted earlier, enforcement is expensive, meaning that it will be very costly to achieve a high probability of imposing a sanction, yet handing down increasingly severe penalties (especially monetary penalties) is relatively costless. However, there are good reasons why a regulator should be cautious about doing this. The prospect of legal errors means that regulatees will forgo activities that are at the borderline of legality, and this is magnified where the penalty is great (Posner 1985, 1206; Yeung 2004, 68). Further, penalties that exceed the means of corporate defendants will flow through to the relatively blameless, such as employees, as the company may be forced to downsize or close (Coffee Jnr 1981, 386–407). Finally, penalties that do not reflect how serious the public sees an offence may undermine the legitimacy of a regulatory regime (Abbot 2009, 35–36; Bilmes and Woodbury 1991, 211) and, as noted above, deterrence already suffers from the criticism that it treats people as a means to an end.

Criminal Law Theory

The RMA's offences are also part of the criminal law, and this engages its own set of theoretical questions. First, picking up where we left off at the end of the previous section, it is necessary to consider what criminal law theory says about the primary rationale for imposing sanctions. Attention then turns to when and how prosecutions should be undertaken, before the section concludes by asking when the criminal law should be employed and how offences should be drafted.

Primary Rationale for Imposing Sanctions: Desert

In contrast to regulatory enforcement theory, criminal law scholars tend to prefer desert as the primary rationale for the imposition of sanctions (see, for instance, Campbell et al. 2019, 23). Unlike deterrence, desert theory is backward-looking and has its origins in retributive philosophy. In particular, it takes the view that penalties should be based on the offence committed, not on the consequences that will come from imposing the penalty.

The purest form of desert theory proceeds on the basis that "we are justified in punishing because and only because offenders deserve it" (Moore 2009, 110), but perhaps the most well-known version is that championed by Von Hirsch (1993). Von Hirsch's theory proceeds on the premise that desert is "an integral part of everyday judgments of praise and blame, which is institutionalised in state punishment to express disapprobation of the conduct and its perpetrators" (Ashworth 2009,

102). This is the censuring aspect of punishment, but it is not enough on its own. The other aspect is hard treatment, which provides a "prudential" incentive not to offend and is required because humans are "moral but fallible creatures" (Von Hirsch and Ashworth 2005, 23). Put another way, "[p]ersons, it is assumed, are neither like angels for whom purely normative appeals would suffice, nor like brutes which could be influenced only by threats" (Von Hirsch and Ashworth 2005, 23).

Censure and hard treatment are intertwined, with the latter being the vehicle by which the former is conveyed. This means that proportionality is critical to desert theory—because people are moral agents, who are expected to respond to the level of censure meted out to them, the penalty imposed must be proportionate to the seriousness of the offending. For instance, a sentence that is disproportionately severe, imposed for (say) deterrence reasons, would be unjust, as the offender would see themselves as receiving a greater amount of censure than they deserve. Proportionality is assessed both ordinally—that is, in terms of the relative seriousness of offences among themselves—and cardinally—that is, for each offence, the penalty should be in proportion to the gravity of the crime involved (Ashworth 2010, 89).

Specific details relating to determining sentences according to desert theory are discussed in Chapter 5, but the foregoing leads Ashworth to conclude that:

> The strengths of desert theory may be recognised in its basis in everyday conceptions of crime and punishment, in its close links with modern liberal political theory, in its insistence that state power be subject to limitations, in its model of individuals as autonomous, choosing beings, and in its protagonists' insistence that sentencing systems should have coherent aims and predictable sentences. (2009, 107)

However, there are both practical and principled criticisms of desert theory. In terms of the former, it is difficult to determine where offences that target very different conduct sit among one another in terms of relative seriousness, and even within an offence why a penalty of a certain size is appropriate. This can mean that there is "considerable scope for inconsistency and the appearance of arbitrariness in setting penalty levels" (Yeung 1999, 457). In terms of principled criticisms, the lack of regard for consequences when imposing a sanction means that, even if a number of negative flow-on effects were to come, such a penalty would still be required.

When and How to Prosecute

Also in contrast to regulatory enforcement scholars, criminal law theorists suggest that there should be limits placed on the use of offences. A sensible starting point for a discussion on this is to recognise that those enforcing the criminal law have considerable discretion. This is broader than simply deciding whether or not to prosecute. If a prosecution is brought, there is also discretion as to who should be charged and for what. Further, there will be discretion as to resolving a prosecution, in terms of whether to engage in plea bargaining (and the terms of any arrangement) or to invite the defendant to take part in a diversionary (that is, out-of-court) scheme. Indeed, such features have led Husak to conclude that where "power *really* is allocated in our criminal justice system today" is with "police and prosecutors" (2008, 21, his emphasis, citations omitted).

Discretion exists in countries like New Zealand because they subscribe to the "opportunity" principle. This principle reflects that even where there is evidence that an offence has been committed, prosecutors have discretion as to whether or not to prosecute. Most famously, it was promoted by a former Attorney-General of the United Kingdom, Sir Hartley Shawcross QC, who in 1951 said that "[i]t has never been the rule in this country … that suspected criminal offences must automatically be the subject of prosecution" (as quoted in Crown Law Office 2013, 5.6). The opportunity principle is often contrasted with the "legality" principle, found in some European jurisdictions, where (at least in theory) the Police must report all offences to the prosecutor who must in each case prosecute (Sanders and Young 2007, 323–324).

It is important, for rule of law reasons, that such discretion is used appropriately. That "like cases be decided alike, the law should apply equally to all, and decision-makers should adhere to procedural requirements of natural justice" are all fundamental aspects of the rule of law (Joseph 2014, 156). Indeed, one of Raz's principles that can be derived from his idea of the rule of law is that "[t]he discretion of the crime-preventing agencies should not be allowed to pervert the law" (Raz 1979, 218). However, as the drafting of offences becomes wider (which is commonly the case in regulatory matters, as discussed next), the more discretion the agency will have and so the more that can go wrong. As Simester and Von Hirsch note, "[w]herever an offence is over-inclusive, so that it requires selective prosecution, the liability of defendants is, in effect, remaindered to the decisions of officials – creating the risk of unfair, inappropriate, or potentially even discriminatory prosecutions" (2011, 206).

One way of ensuring that discretion is being used properly, and the public know what to expect from an enforcer, is by implementing constraints. As Beazley and Pulsford note, when discussing the related rule of law value of certainty, "[t]he law cannot be clear and predictable, and its operation certain, if the executive and the judiciary have a discretion at large to do as they please" (2015, 160). One type of constraint is a requirement that those applying sanctions must publish an enforcement policy. This has been the approach taken by the Australian Law Reform Commission, which has recommended that "[r]egulators who administer legislation under which criminal, civil or administrative penalties may be imposed or arise should develop and publish enforcement guidelines setting out their enforcement approach", and that this should cover (unless clearly inappropriate) the types of action available, the principles behind each action, the criteria involved in the decision to pursue one or more of the actions and the regulator's relationship with other regulators and enforcement agencies (2002, 388–389).

Policies or guidelines on how to exercise discretion do not guarantee that it will be used properly. As such, it also important to identify when discretion has been used inappropriately, and this can be achieved by making those responsible for applying offences accountable. This includes not only decisions to prosecute, but also decisions not to prosecute (Duff et al. 2007, 189). Accountability—the "duty to give account for one's actions to some other person or body"—can take a number of different forms, such as political, administrative, financial, legal and professional (Scott 2000, 40). For instance, there could be an independent inspectorate (for enforcement action generally), oversight by the Solicitor-General (for criminal proceedings generally) and review from the courts (for specific prosecutions brought or not brought). It must be observed, however, that in terms of judicial review by the courts, judges are generally wary about intervening in the exercise of prosecutorial discretion (Finn and Mathias 2019, 66–68).

Finally, it is sometimes argued that where there is an accountability gap this can be filled by private prosecutions. These have been said, at least historically, to be "a useful constitutional safeguard against capricious, corrupt or biased failure or refusal of those authorities to prosecute offenders against the criminal law" (*Gouriet v Union of Post Office Workers* [1978] AC 435 at 498). However, in modern times such prosecutions are rarely brought due to the cost and energy involved, meaning that this form of accountability can be more apparent than real (Sanders and Young 2007, 379).

Employment and Drafting of Offences

As Ball and Friedman have observed, the term "use" includes not just the application of offences to transgressors (that is, prosecution), but also the employment (and drafting) of offences (1965, 199), and criminal law theory also urges for limits to be imposed on this front. This is not a question that much concerns regulatory scholars, the typical view being simply that the criminal law is "just another state regulatory tool", its purpose being to "visit punishments such as imprisonment or fines in situations where other forms of disincentive make for an insufficient deterrent" (Simester and Von Hirsch 2011, 4, citing Posner 1985 as an example of such an approach). At the other end of the spectrum, a pure desert theorist would suggest that the criminal law "exists *in order* to punish people for their culpable misconduct" and therefore "one should criminalise, and punish, *whenever* there is a culpable wrong, *and not otherwise*" (Simester and Von Hirsch 2011, 4, their emphasis, citing Moore 1997 as an example of such an approach).

While criminal law scholars generally reject the former approach (see, for instance, Duff 2014, 223), most would also not go as far as the latter end of the spectrum. For instance, on Simester and Von Hirsch's account the criminal law "is a regulatory tool for influencing behaviour, and in some respects no more than that; but it is a special kind of tool" (2011, 4). In particular, it is:

> ... distinctive because of its moral voice. It removes specified activities from the permissible and punishes individuals who venture or stray into its realm. It is a complex, authoritative, censuring device. Conduct is deemed through its criminalisation to be, and is subsequently punished as, *wrongful* behaviour that warrants blame. (2011, 19, their emphasis)

On such an approach, "a core element of criminal law, from a normative point of view, is that the criminal sanction should be reserved for substantial wrongdoing", the two "main dimensions" of which are "harm and culpability" (Ashworth 2000, 240). The touchstones of wrongs, and culpability, and the relationship between them, is, therefore, a useful framework for an analysis of when the criminal law should be employed. It is necessary to start with the employment of the criminal law generally, before looking at the specific challenges posed by a regulatory context.

Ashworth's formulation, referred to in the paragraph above, refers to wrongdoing by reference to harm and culpability. But before harm and culpability can be discussed, a point needs to be made about the concept of "wrongdoing" itself. This is that the criminal law is not concerned with all types of wrongs. For instance, private wrongs are generally considered the domain of the civil law, if they are to be regulated at all. Rather, the criminal law is concerned with public wrongs, although this does not necessarily mean wrongs against the public (although in many instances these will be included). Instead, the criminal law is concerned with public wrongs in the sense that they are wrongs that "the community is responsible for punishing" (Lamond 2007, 614). Duff and Marshall refer to this as "shared wrongs", justifying it as follows:

> ...we share *in* the very wrong that [the victim] has suffered: it is not 'our' wrong *instead* of hers; it is 'our' wrong *because* it is a wrong done to her,

> as one of us-as a fellow member of our community whose identity and whose good is found within that community. (1998, 21, their emphasis)

Turning then to harm, this is most commonly reflected in the "Harm Principle". This principle was first described by Mill in an exclusionary manner—"[t]he *only* purpose for which power can be rightfully exercised over any member of a civilized community, against his will, is to prevent harm to others" (1859, 22, emphasis added)—but more recently has been considered a positive justification for criminalisation:

> It is always a good reason in support of penal legislation that it would be effective in preventing (eliminating, reducing) harm to persons other than the actor (the one prohibited from acting) and that there is no other means that is equally effective at no greater cost to other values. (Feinberg 1984, 26)

Harm is generally considered to be the "thwarting, setting back, or defeating of an interest" (Feinberg 1984, 33). What interests are included in this is a matter of debate, but as Lin notes "certain interests are so indisputably critical that setbacks to them are commonly recognized as harm, regardless of the social norms applied … [and this] … may include physical well-being, emotional well-being, and property interests" (Lin 2006, 938–939). It generally also includes matters such as conduct that creates or causes the risk of harm to others (Duff 2007, 126; Horder 2012).

This leaves culpability. As Simester and Von Hirsch observe, the need for wrongfulness is sometimes confused with the need for culpability elements, but both are in fact required if one is to be blamed (2011, 23). For instance, a person can do something "wrong" (for instance, push another person over), but not be culpable (for instance, because they fell by accident into the other person). Equally, one can perform an act culpably, but if it was not (a public) wrong in the first place then they should not be blamed. The "centrality of the culpability requirement" to serious wrongdoing is said to be "surely part of the essence of the criminal law" (Ashworth 2000, 240) and stems from the requirement of fairness, in that it is "not only unfair to censure people who are not culpable, but also unfair to punish them for the offence" (Ashworth and Horder 2013, 162). Indeed, as Sayre argued nearly 100 years ago:

> The moral obloquy and the social disgrace incident to criminal conviction are whips which lend effective power to the administration of criminal law. When the law begins to permit convictions for serious offenses of men who are morally innocent and free from fault, who may even be respected and useful members of the community, its restraining power becomes undermined. Once it becomes respectable to be convicted the vitality of the criminal law has been sapped. (1933, 79–80)

Returning then to the quote of Ashworth that this discussion began with, the final aspect to discuss is his qualification that the wrongdoing be "substantial". There are (at least) two different ways to consider this. One is from a purely practical perspective—criminal justice processes are time consuming and costly, and therefore should be reserved for conduct that is serious enough to justify this expense. But the "substantial" qualification can also, and more importantly, be seen as reflecting a "minimalist approach" to the question of criminalisation, which reflects:

> a) the principle of respect for human rights, b) the right not to be subjected to State punishment, c) the principle that the criminal law should not be invoked unless other techniques are inappropriate, and d) the principle that conduct should not be criminalized if the effects of doing so would be as bad as, or worse than, not doing so. (Ashworth and Horder 2013, 31)

If the criminal law should be reserved for "substantial wrongdoing", does this mean that it should not be used for regulatory breaches? This argument is sometimes made, on the basis that such breaches are often said to be "morally neutral". But as Green notes, one must be clear as to what is meant by moral neutrality, as this can refer to a variety of things: that the conduct in question is not a public wrong, that the conduct is not harmful enough to justify using the criminal law or that offenders who are not culpable are liable (1997, 1537). It is necessary, therefore, to return to the three concepts of wrongs, harmfulness and culpability and consider how, if at all, employing the criminal law for regulatory breaches can be justified.

Looking first at wrongs, it is sometimes said that regulatory breaches do not satisfy this criterion because the underlying conduct is *mala prohibita*—that is, the breaches are not wrong prior to or independent of the law. This is then contrasted with conduct that is *mala in se* (intrinsically wrongful or wrong in and of itself) and which is said to be the proper domain of the criminal law. However, establishing whether certain conduct is mala prohibita or mala in se is often difficult, given the differing values held both within and between societies. This is particularly the case when in some instances conduct that traditionally would have been considered "only regulatory" generates such public concern that it becomes "truly criminal". Indeed, many offences that are referred to as mala prohibita are at least "hybrids" (as termed by Husak 2005, 75), in that they "involve a more or less artificial, stipulative determination of a genuine *malum in se*" (Duff 2002, 102).

But even where the *mala prohibita* label is accurate, this does not necessarily mean that the conduct is not sufficiently wrong to justify the employment of the criminal law. This is because such offences can be supported on the basis that, so long as the regulatory rules are properly made and serve some societal good, it is a wrong to breach them (Duff 2010, 89). Duff justifies this on the basis that it is part of our "civic duty" to comply with these rules (2007, 166–174), while for Lamond it is wrong because to breach such rules shows "disrespect" for schemes of conduct that have the effect of co-ordinating risk reduction or the promotion of certain goods (2007, 630). On the other hand, Simester and Von Hirsch consider such justifications unnecessary, arguing that "[i]f the state's exercise of authority [in regulating an action] was justified, then [that action] is already wrongful and, post-legally, D ought not to do it" (2011, 28–29). Whatever the rationale, the critical point, as Duff notes, is that the conduct must still be "*pre-criminally* wrongful" (2014, 219, his emphasis)—that is, the conduct does not need to be wrongful independent of the law as a whole, but it must be wrongful before it is criminalised.

In terms of harm and culpability, employing the criminal law for regulatory breaches is sometimes questioned because conduct that does not cause actual (or even the risk of) harm, and that is committed unintentionally, is often criminalised. However, at the same time, some regulatory breaches are very harmful and are committed culpably, clearly justifying the use of the criminal law. By way of example, in (at least) modern times there would seem to be little doubt that the intentional dumping of toxic chemicals into an environmentally significant waterway should be an offence. The problem is actually, therefore, when the two extremes (and everything in between) are bundled up together—for instance, it is common to find a very wide range of conduct criminalised in one offence, from cases where there is no harm (and no risk of harm), and which can be committed accidentally, up to significant actual harm caused intentionally (O'Hear 2004, 147–148).

One issue that arises from this is related to the earlier discussion on when and how to prosecute—"[l]egislative crime definition has a natural tendency to become, in practice, prosecutorial crime definition, as legislatures define broad nominal liability rules, leaving prosecutors to determine what behavior actually leads to conviction and punishment" (Stuntz 2001, 578; see also Stuntz 2011, 4). But more broadly, from a normative criminality perspective, if the criminal sanction is properly reserved for substantial wrongdoing, assessed predominantly by reference to harm and culpability, then it follows that this should be reflected in the drafting of offences. In other words, a specific offence should only prohibit sufficiently harmful conduct (or conduct that risks harm) that is committed culpably.

In general, the need for harm is tracked by the *actus reus* and the need for culpability by the *mens rea* element (Simester et al. 2013, 18), so it is worth considering the approach, in regulatory matters, often taken to each of these matters. Starting with the *actus reus* of an offence, there is a tendency in regulatory matters to make this very broad. An element of this breadth is inescapable, in that it is not possible to specifically provide for every possible instance of harmful conduct. But, as noted above, the problem is that often little attempt is made to distinguish between harmful and non-harmful conduct, with all such behaviour included and the task of determining between the two left to the enforcer. Further, the tendency is always to expand (rather than contract) the *actus reus*, as governments become increasingly concerned that people might escape liability through a "loophole". This undoubtedly provides flexibility for the regulator, which feeds into the enforcement strategies adopted, but runs the risk of infringing on the principle of legality, which holds that "the legislator should prescribe the criminalised behaviour as precisely as possible" (Faure 2017a, 272).

There are two related points to this. One is the vague manner in which regulatory rules are often drafted (as referred to in the regulatory enforcement section of this chapter), and to which offences are attached, which runs contrary to the requirement that "legislation be certain, and that liability under the criminal law be predictable and ascertainable in advance" (Lees 2015, 27). The other is the overall liability model, whereby (for instance) polluting per se is not criminalised, but rather the offence is one of polluting without a licence (or outside the terms of the licence). This has been described as the administrative dependence of environmental criminal law (Faure 2017b, 327), a consequence of which is that:

> administrative authorities…receive wide powers to determine the punishable nature of certain polluting acts. Indeed, administrative authorities define the conditions of a permit and with that they effectively also determine the conditions for criminal liability. (Faure 2017b, 328).

There is also a temptation in regulatory matters to eliminate the need for *mens rea* from an offence, thereby providing for liability without fault. In New Zealand, common law liability without fault is typically divided into two forms—strict liability and absolute liability. Strict liability means that the prosecution does not need to prove any mental element, but the defendant will be exculpated if it can show, on the balance of probabilities, a total absence of fault (see *Civil Aviation Department v MacKenzie* [1983] NZLR 78 (CA) and *Millar v Ministry of Transport* [1986] 1 NZLR 660 (CA)). For absolute liability, no such defence exists. Finally, a third option is for the statute to explicitly state that the prosecution does not need to prove intention. In these circumstances, the legislation will usually provide what defences are available, to the exclusion of the common law defence.

Liability without fault is often justified based on its perceived instrumental benefits. For instance, Simester identifies five such arguments in favour of it: that the cost should, prima facie, be reduced because there are less elements for the prosecution to prove; that outcomes should be more accurate as there are less elements that could be decided wrongly; that it may assist with prosecuting companies, as there is no need to find a person whose state of mind can be attributed to the company; that it will increase the deterrent effect of the prohibition because people will take more care if they know they can be prosecuted without being at fault; and that it might be more efficient, in terms of allocating costs, than tort law (2005, 25–30). However, he also goes on to note that at least some of these benefits can be overstated: in terms of cost and accuracy, culpability is still relevant to sentencing, and so the argument about it, and the cost and accuracy of the determination, may simply shift to a different stage in proceedings, and it is also arguable that liability without fault may in fact be less of a deterrent, as the defendant, assuming it wants to remain in the industry and it is not cost effective to reduce the chance of offending to zero, will be unaware what level of precaution to aim for (Simester 2005, 30–33).

Further, the practical benefits do not detract from the point that, because of the centrality of culpability to the criminal law, liability without fault "is not to be welcomed, in general. It should be regarded as exceptional and in need of strong justification, particularly where the offence is serious" (Ashworth and Zedner 2008, 31–33). This position is, of course, tempered somewhat when there is a total absence of fault defence, but where the defence instead comes from a statutory provision its legitimacy will depend on the fairness of the defence(s) provided. Put another way, a defence that can only be relied on in extremely limited situations will mean that the offence is closer to absolute liability than strict liability. This again relates back to the application of offences, with Situ and Emmons observing that, due to the prevalence of liability without fault, "discretion in prosecuting environmental offenders raises more serious questions about justice and fairness than it normally does in prosecuting conventional criminals" (2000, 43).

The Framework for Assessment

The foregoing review of regulatory enforcement and criminal law theory has identified two key areas of tension. First, regulatory enforcement theorists and criminal law scholars disagree over the primary rationale for imposing penalties—that is, what makes an offence *effective*. The former considers that an effective sanction is one that deters others from offending, with the latter suggesting that an effective sanction is one that is proportionate to the offence committed (desert theory). Second, there is also disagreement over whether offences need to be applied *appropriately*. While regulatory enforcement theory has few concerns as to how agencies are applying offences in response to non-compliance—generally, enforcers should choose whichever tool will maximise compliance and apply the tool in the way that will support this—criminal law scholars are much more worried about what prosecutions are undertaken and the manner in which they are conducted.

In terms of these two points of contention, the case study on the RMA's offences in Part III of this book proceeds on the following basis. No normative position is taken on the rationale for imposing sanctions, given the availability of two viable options (deterrence and desert). However, this makes it necessary to ask why the local authorities and the courts say they are prosecuting and imposing sentences (respectively). This is the first task of Chapter 4, followed by an assessment of whether the offences are achieving this goal. After this, Chapter 5 looks at what is *in fact* happening when it comes to the imposition of sanctions.

On the other hand, when it comes to considering how local authorities are applying the offences to those not complying with the rules, the criminal law approach—that the offences must be applied appropriately—is adopted. It is suggested that this is the only justifiable position given the serious consequences that can come from a successful RMA prosecution, including a loss of liberty and a conviction. Further, there is no principled reason for treating the RMA's offences in a different way to how offences in a mainstream criminal statute would be treated (see generally Ashworth and Redmayne 2005, 171)—complaints would rightly be made if the Police were using the offences in the Crimes Act 1961 in an inappropriate manner, so why should things be any different just because it is a local authority and the RMA? Chapter 6 therefore assesses how local authorities are using the offences against well-established criminal law values such as equality of treatment and fairness.

This is not the end of the matter, however. Just because the RMA's offences are formally criminal, this is not to say that the employment of the criminal law and the way the offences are drafted complies with ideals of normative criminality—that is, only proscribing conduct that is sufficiently wrong, harmful and culpable. Whether or not this is the case which is addressed in Chapter 7, along with an analysis of the RMA's penalties and institutional arrangements for prosecuting, in order to determine whether these features of the Act are driving the approach that local authorities are taking to the offences.

References

Abbot, Carolyn. 2009. *Enforcing Pollution Control Regulation*. Portland: Hart Publishing.

Almond, Paul, and Sarah Colover. 2012. Communication and Social Regulation: The Criminalization of Work-Related Death. *The British Journal of Criminology* 52 (5): 997–1016. https://doi.org/10.1093/bjc/azs038.

Ashworth, Andrew. 2000. Is the Criminal Law a Lost Cause? *Law Quarterly Review* 116: 225–256.

Ashworth, Andrew. 2009. Desert. In *Principled Sentencing: Readings on Theory and Policy*, ed. Andreas Von Hirsch, Andrew Ashworth, and Julian Roberts, 102–109. Portland: Hart Publishing.

Ashworth, Andrew. 2010. *Sentencing and Criminal Justice*, 5th ed. Cambridge: Cambridge University Press.

Ashworth, Andrew, and Jeremy Horder. 2013. *Principles of Criminal Law*, 7th ed. Oxford: Oxford University Press.

Ashworth, Andrew, and Mike Redmayne. 2005. *The Criminal Process*, 3rd ed. New York: Oxford University Press.

Ashworth, Andrew, and Lucia Zedner. 2008. Defending the Criminal Law: Reflections on the Changing Character of Crime, Procedure, and Sanctions. *Criminal Law and Philosophy* 2: 21–51.

Australian Law Reform Commission. 2002. Principled Regulation: Federal Civil and Administrative Penalties in Australia: ALRC R95.

Ayres, Ian, and John Braithwaite. 1992. *Responsive Regulation: Transcending the Deregulation Debate*. New York: Oxford University Press.

Baldwin, Robert. 1990. Why Rules Don't Work. *Modern Law Review* 53 (3): 321–337. https://doi.org/10.1111/j.1468-2230.1990.tb01815.x.

Baldwin, Robert, and Julia Black. 2008. Really Responsive Regulation. *Modern Law Review* 71 (1): 59–94.

Ball, Harry, and Lawrence Friedman. 1965. The Use of Criminal Sanctions in the Enforcement of Economic Legislation: A Sociological View. *Stanford Law Review* 17 (2): 197–223.

Beazley, M., and Myles Pulsford. 2015. Discretion and the Rule of Law in the Criminal Justice System. *Australian Law Journal* 89 (3): 158–174.

Beccaria, Cesare. 1764. *On Crimes and Punishments*.

Becker, Gary. 1968. Crime and Punishment: An Economic Approach. *Journal of Political Economy* 76 (2): 169–217.

Bentham, Jeremy. 1789. *An Introduction to the Principles of Morals and Legislation*.

Bilmes, Jack, and John Woodbury. 1991. Deterrence and Justice: Setting Civil Penalties in the Federal Trade Commission. *Research in Law and Economics* 14: 191–216.

Black, Julia, and Robert Baldwin. 2010. Really Responsive Risk-Based Regulation. *Law & Policy* 32 (2): 181–213.

Black, Julia, and Robert Baldwin. 2012. When Risk-Based Regulation Aims Low: Approaches and Challenges. *Regulation & Governance* 6 (1): 2–22. https://doi.org/10.1111/j.1748-5991.2011.01124.x.

Braithwaite, John. 2002. *Restorative Justice & Responsive Regulation*. Oxford: Oxford University Press.

Braithwaite, John, and Gilbert Geis. 1982. On Theory and Action for Corporate Crime Control. *Crime & Delinquency* 28 (2): 292–314. https://doi.org/10.1177/001112878202800207.

Brown, R.M. 1992. Administrative and Criminal Penalties in the Enforcement of Occupational Health and Safety Legislation. *Osgoode Hall Law Journal* 30 (3): 691–735.

Burby, Raymond, and Robert Paterson. 1993. Improving Compliance with State Environmental Regulations. *Journal of Policy Analysis and Management* 12 (4): 753–772.

Campbell, Liz, Andrew Ashworth, and Mike Redmayne. 2019. *The Criminal Process*, 5th ed. Oxford: Oxford University Press.

Coffee Jnr, John C. 1981. "No Soul to Damn: No Body to Kick": An Unscandalized Inquiry into the Problem of Corporate Punishment. *Michigan Law Review* 79 (3): 386–459.

Cooter, Robert. 1984. Prices and Sanctions. *Columbia Law Review* 84 (6): 1523–1560.

Crown Law Office. 2013. Solicitor-General's Prosecution Guidelines.

Diver, Colin. 1980. A Theory of Regulatory Enforcement. *Public Policy* 28 (3): 257–299.

Duff, R.A. 2002. Crime, Prohibition, and Punishment. *Journal of Applied Philosophy* 19 (2): 97–108. https://doi.org/10.1111/1468-5930.00207.

Duff, R.A. 2007. *Answering for Crime: Responsibility and Liability in the Criminal Law*. Oxford: Hart Publishing.

Duff, R.A. 2010. Perversions and Subversions of Criminal Law. In *The Boundaries of the Criminal Law*, ed. R.A. Duff, Lindsay Farmer, Sandra Marshall, Massimo Renzo, and Victor Tadros, 88–112. New York: Oxford University Press.

Duff, R.A. 2014. Towards a Modest Legal Moralism. *Criminal Law and Philosophy* 8 (1): 217–235. https://doi.org/10.1007/s11572-012-9191-8.

Duff, R.A., L. Farmer, S. Marshall, and V. Tadros. 2007. *Trial on Trial: Volume 3: Towards a Normative Theory of the Criminal Trial*. Oxford: Hart Publishing.

Faure, Michael. 2017a. The Evolution of Environmental Criminal Law in Europe: A Comparative Analysis. In *Environmental Crime in Europe*, ed. Andrew Farmer, Michael Faure, and Grazia Maria Veronica. Vagliasindi, 267–318. Oxford: Hart Publishing.

Faure, Michael. 2017b. The Revolution in Environmental Criminal Law in Europe. *Virginia Environmental Law Journal* 35 (2): 321–356.

Faure, Michael, and Katarina Svatikova. 2012. Criminal or Administrative Law to Protect the Environment? Evidence from Western Europe. *Journal of Environmental Law* 24 (2): 253–286. https://doi.org/10.1093/jel/eqs005.

Feinberg, Joel. 1984. *The Moral Limits of the Crimial Law, Vol. 1: Harm to Others*. New York: Oxford University Press.

Finn, Jeremy, and Don Mathias. 2019. *Criminal Procedure in New Zealand*, 3rd ed. Wellington: Thomson Reuters.

Garoupa, Nuno. 1997. The Theory of Optimal Law Enforcement. *Journal of Economic Surveys* 11 (3): 267–295. https://doi.org/10.1111/1467-6419.00034.

Garoupa, Nuno, Anthony Ogus, and Andrew Sanders. 2011. The Investigation and Prosecution of Regulatory Offences: Is There an Economic Case for Integration? *Cambridge Law Journal* 70 (1): 229–259.

Genn, Hazel. 1993. Business Responses to the Regulation of Health and Safety in England. *Law & Policy* 15 (3): 219–233. https://doi.org/10.1111/j.1467-9930.1993.tb00104.x.

Gray, Wayne, and John Scholz. 1993. Does Regulatory Enforcement Work? A Panel Analysis of OSHA Enforcement. *Law & Society Review* 27 (1): 177–213.

Green, Stuart. 1997. Why It's a Crime to Tear the Tag Off a Mattress: Overcriminalization and the Moral Content of Regulatory Offenses. *Emory Law Journal* 46 (4): 1533–1615.

Gunningham, Neil. 2010. Enforcement and Compliance Strategies. In *The Oxford Handbook of Regulation*, ed. Robert Baldwin, Martin Cave, and Martin Lodge, 120–145. New York: Oxford University Press.

Gunningham, Neil. 2011. Enforcing Environmental Regulation. *Journal of Environmental Law* 23 (2): 169–201.

Gunningham, Neil, Peter Grabosky, and Darren Sinclair. 1998. *Smart Regulation: Designing Environmental Policy*. Oxford: Clarendon Press.

Gunningham, Neil, and Richard Johnstone. 1999. *Regulating Workplace Safety: System and Sanctions*. Oxford: Oxford University Press.

Hawkins, Keith. 1984. *Environment and Enforcement: Regulation and the Social Definition of Pollution*. Oxford: Clarendon Press.

Hawkins, Keith. 2002. *Law as Last Resort: Prosecution Decision-Making in a Regulatory Agency*. Oxford: Oxford University Press.

Hawkins, Keith. 2013. Enforcing Regulation: Robert Kagan's Contribution—And Some Questions. *Law & Social Inquiry* 38 (4): 950–972. https://doi.org/10.1111/lsi.12005.

Horder, Jeremy. 2012. Harmless Wrongdoing and the Anticipatory Perspective on Criminalisation. In *Seeking Security: Pre-empting the Commission of Criminal Harms*, ed. G.R. Sullivan and Ian Dennis, 79–102. Oxford: Hart Publishing.

Husak, Douglas. 2005. Malum Prohibitum *and Retributivism*. In *Defining Crimes: Essays on the Special Part of the Criminal Law*, ed. R.A. Duff and Stuart Green, 65–90. Oxford: Oxford University Press.

Husak, Douglas. 2008. *Overcriminalization: The Limits of the Criminal Law*. Oxford: Oxford University Press.

Hutter, Bridget. 1997. *Compliance: Regulation and Environment*. Oxford: Clarendon Press.

Joseph, Philip. 2014. *Constitutional and Administrative Law in New Zealand*, 4th ed. Wellington: Brookers.

Kagan, Robert, and John Scholz. 1984. The "Criminology of the Corporation" and Regulatory Enforcement Strategies. In *Enforcing Regulation*, ed. Keith Hawkins and John Thomas. Boston: Kluwer Nijhoff Publishing.

Lamond, Grant. 2007. What Is a Crime? *Oxford Journal of Legal Studies* 27 (4): 609–632.

Landes, William, and Richard Posner. 1975. The Private Enforcement of Law. *The Journal of Legal Studies* 4 (1): 1–46. https://doi.org/10.1086/467524.

Lees, Emma. 2015. *Interpreting Environmental Offences: The Need for Certainty*. Oxford: Hart Publishing.

Lin, Albert. 2006. The Unifying Role of Harm in Environmental Law. *Wisconsin Law Review* 2006 (3): 897–985.

Marshall, Sandra, and R.A. Duff. 1998. Criminalization and Sharing Wrongs. *The Canadian Journal of Law and Jurisprudence* 11 (1): 7–22. https://doi.org/10.1017/S0841820900001661.

May, Peter. 2004. Compliance Motivations: Affirmative and Negative Bases. *Law & Society Review* 38 (1): 41–68. https://doi.org/10.1111/j.0023-9216.2004.03801002.x.

McKean, Roland. 1980. Enforcement Costs in Environmental and Safety Regulation. *Policy Analysis* 6 (3): 269–289.

Mill, John Stuart. 1859. *On Liberty*. London: Parker.

Moore, Michael. 1997. *Placing Blame: A General Theory of the Criminal Law*. Oxford: Clarendon Press.

Moore, Michael. 2009. The Moral Worth of Retribution. In *Principled Sentencing: Readings on Theory and Policy*, ed. Andreas Von Hirsch, Andrew Ashworth, and Julian Roberts, 110–114. Portland: Hart Publishing.

Nurse, Angus. 2016. *An Introduction to Green Criminology and Environmental Justice*. London: Sage.

O'Hear, Michael. 2004. Sentencing the Green-Collar Offender: Punishment, Culpability, and Environmental Crime. *The Journal of Criminal Law and Criminology* 95 (1): 133–276.

Ogus, Anthony. 1992. Regulatory Law: Some Lessons from the Past. *Legal Studies* 12 (1): 1–19. https://doi.org/10.1111/j.1748-121X.1992.tb00453.x.

Ogus, Anthony, and Carolyn Abbot. 2002. Sanctions for Pollution: Do We Have the Right Regime? *Journal of Environmental Law* 14 (3): 283–298.

Organisation for Economic Co-operation and Development. 2010. Risk and Regulatory Policy: Improving the Governance of Risk. OECD Reviews of Regulatory Reform.

Organisation for Economic Co-operation and Development. 2014. Regulatory Enforcement and Inspections: OECD Best Practice Principles for Regulatory Policy.

Paternoster, Raymond. 2009. Deterrence and Rational Choice Theories. In *21st Century Criminology: A Reference Handbook*, ed. J Mitchell Miller. Thousand Oaks: Sage.

Paternoster, Raymond. 2010. How Much Do We Really Know About Criminal Deterrence? *The Journal of Criminal Law and Criminology* 100 (3): 765–824.

Polinsky, A. Mitchell., and Steven Shavell. 2000. The Economic Theory of Public Enforcement of Law. *Journal of Economic Literature* 38 (1): 45–76.

Polinsky, A. Mitchell, and Steven Shavell. 2007. The Theory of Public Enforcement of Law. In *Handbook of Law and Economics*, ed. A. Mitchell Polinsky and Steven Shavell, 403–454. Amsterdam: Elsevier.

Posner, Richard. 1985. An Economic Theory of the Criminal Law. *Columbia Law Review* 85 (6): 1193–1231.

Raz, Joseph. 1979. *The Authority of Law: Essays on Law and Morality*. Oxford: Clarendon Press.

Reiss Jnr, Albert J. 1984. Selecting Strategies of Social Control Over Organisational Life. In *Enforcing Regulation*, ed. Keith Hawkins and John Thomas. Boston: Kluwer-Nijhoff Publishing.

Roberts, Julian, and Andrew Ashworth. 2009. Deterrence. In *Principled Sentencing: Readings on Theory and Policy*, ed. Andreas Von Hirsch, Andrew Ashworth, and Julian Roberts, 39–48. Portland: Hart Publishing.

Sanders, Andrew, and Richard Young. 2007. *Criminal Justice*, 3rd ed. Oxford: Oxford University Press.

Sayre, Francis. 1933. Public Welfare Offenses. *Columbia Law Review* 33 (1): 55–88.

Scholz, John. 1984. Cooperation, Deterrence, and the Ecology of Regulatory Enforcement. *Law & Society Review* 18 (2): 179–224.

Scott, Colin. 2000. Accountability in the Regulatory State. *Journal of Law and Society* 27 (1): 38–60. https://doi.org/10.1111/1467-6478.00146.

Selznick, Philip. 1985. Focusing Organisational Research on Regulation. In *Regulatory Policy and the Social Sciences*, ed. Roger Noll, 363–364. Berkley: University of California Press.

Silberman, Jon. 2000. Does Environmental Deterrence Work? Evidence and Experience Says Yes, But We Need to Understand How and Why. *The Environmental Law Reporter* 30 (7): 10523–10536.

Simester, A.P. 2005. Is Strict Liability Always Wrong. In *Appraising Strict Liability*, ed. A.P. Simester, 21–50. Oxford: Oxford University Press.

Simester, A.P., J.R. Spencer, G.R. Sullivan, and G.J. Virgo. 2013. *Simester and Sullivan's Criminal Law: Theory and Doctrine*, 5th ed. Oxford: Hart Publishing.

Simester, A.P., and Andreas Von Hirsch. 2011. *Crimes, Harms, and Wrongs: On the Principles of Criminalisation.* Oxford: Hart Publishing.

Situ, Yingyi, and David Emmons. 2000. *Environmental Crime: The Criminal Justice System's Role in Protecting the Environment*. Thousand Oaks: Sage.

Stretesky, Paul B. 2006. Corporate Self-Policing and the Environment. *Criminology (Beverly Hills)* 44 (3): 671–708. https://doi.org/10.1111/j.1745-9125.2006.00060.x.

Stuntz, William J. 2001. The Pathological Politics of Criminal Law. *Michigan Law Review* 100 (3): 505–600. https://doi.org/10.2307/1290411.

Stuntz, William J. 2011. *The Collapse of American Criminal Justice*. Cambridge: Harvard University Press.

Sutherland, Edwin. 1983. *White Collar Crime: The Uncut Version.* New Haven: Yale University Press.

Von Hirsch, Andreas. 1993. *Censure and sanctions*. New York: Clarendon Press.

Von Hirsch, Andreas, and Andrew Ashworth. 2005. The Justification for Punishment's Existence: Censure and Prevention. In *Proportionate Sentencing: Exploring the Principles*, ed. Andreas Von Hirsch and Andrew Ashworth, 12–34. New York: Oxford University Press.

Von Hirsch, Andreas, A. Bottoms, E. Burney, and P.-O. Wikstrom. 1999. *Criminal Deterrence and Sentence Severity: An Analysis of Recent Research.* Oxford: Hart Publishing.

Wells, Celia. 1993. *Corporations and Criminal Responsibility.* Oxford: Clarendon Press.

White, Rob. 2017. Reparative Justice, Environmental Crime and Penalties for the Powerful. *Crime, Law and Social Change* 67 (2): 117–132. https://doi.org/10.1007/s10611-016-9635-5.

Windholz, Eric. 2017. *Governing Through Regulation: Public Policy, Regulation and the Law*. New York: Routledge.

Winter, Søren., and Peter May. 2001. Motivation for Compliance with Environmental Regulations. *Journal of Policy Analysis and Management* 20 (4): 675–698. https://doi.org/10.1002/pam.1023.

Yeung, Karen. 1999. Quantifying Regulatory Penalties: Australian Competition Law Penalties in Perspective. *Melbourne University Law Review* 23 (2): 440–475.

Yeung, Karen. 2004. *Securing Compliance: A Principled Approach*. Oxford: Hart Publishing.

Part III

The Case Study

4

Are the RMA's Offences Effective?

Having established the framework against which the RMA's offences are to be assessed, it is time to now consider the first criterion, namely whether the offences are effective. However, as noted at the end of the previous chapter, the answer to this question will depend on the reason why offenders are being prosecuted and penalised, and there are different views held as to the proper rationale for this. On the one hand, because the RMA's offences are part of the criminal law, it may be that penalties are being sought and imposed because they are deserved. On the other hand, since the offences are also part of a regulatory regime, it could be that penalties are being sought and imposed to deter other, would-be, offenders. Indeed, it may even be that some other rationale for sanctioning has been adopted by local authorities and the courts. This is, therefore, the first matter to which we turn our attention in this chapter, starting with the courts before moving on to local authorities. After this, the offences are assessed according to the identified rationale.

M. Wright, *Responding to Environmental Crimes*, Palgrave Studies in Green Criminology, https://doi.org/10.1007/978-3-030-89250-0_4

Desert, Deterrence or Something Else?

In determining the courts' rationale for imposing penalties, the natural point of departure is the case of Machinery Movers Ltd v Auckland Regional Council [1994] 1 NZLR 492 (HC). This was the first RMA sentencing decision to reach the High Court on appeal, coming nearly two years after the Act came into force. In this case, the defendant had been sentenced at first instance to a penalty of $25,000 for discharging contaminants onto land in circumstances where they may enter water (a contravention of section 15(1)(b) of the RMA). Given the absence of any RMA-specific statutory sentencing principles, a "Full Court" (two judges) was convened for the appeal to consider the proper approach to these cases.

After setting out the facts of the case and summarising the first instance decision, the High Court began its observations by drawing a line under the sentencing decisions in the previous legislation. The Court noted that the RMA is "informed by a wholly different environmental philosophy, which places far greater emphasis on environmental protection and introduces a much more stringent regime of penalties and punishment" (*Machinery Movers Ltd*, 499). It then went on to suggest that there is clearly a role for deterrence in sentencing RMA offenders, stating that:

> An increase of one third in the maximum fine, the inclusion of imprisonment as a sentencing option, and the addition of director's liability signify an evident legislative dissatisfaction with the level of penalties imposed under the 1967 Act. In combination, these changes constitute a *clear legislative direction to the Courts to ensure that higher penalties are imposed which will have a significant deterrent quality*. If fines are too low, they will be regarded as a minor licence fee for offending and convey the idea that the law may be broken with relative impunity. (500, emphasis added)

However, the Court then observed that other purposes of sentencing were also relevant. In particular, the Judges observed that the "much wider sentencing powers given by the RMA…was a deliberate move to encourage a flexible and innovative approach to sentencing, which seeks not only to punish offenders but also to achieve economic and educative goals" (*Machinery Movers Ltd*, 501).

These latter comments would turn out to be consistent with the subsequent enactment of the Sentencing Act 2002, which applies to all criminal offending and provides a range of purposes for which a court "may sentence or otherwise deal with an offender" (section 7(1)). These include purposes closely linked to desert (section 7(1)(a) and (b)) and deterrence (section 7(1)(f)), as well as other rationales such as protecting the community from the offender (section 7(1)(g)). Indeed, to further complicate matters, a court can adopt a combination of two or more of the purposes (section 7(1)(i)).

However, in a decision issued shortly after the Sentencing Act 2002 came into force, *Waitakere City Council v Gionis* DC Auckland CRN1090034293, 17 December 2002, the sentencing Judge noted that deterrence was—and remains—the primary reason for imposing penalties under the RMA. In particular, it was observed that:

> Consistent with the greater focus on environmental protection the cases show that deterrence is usually one of the uppermost factors in the mind of the Court in sentencing in environmental cases. The position has not been altered by the advent of the Sentencing Act 2002 which provides the Court with a number of alternative purposes and recognises that in different types of cases different purposes will be appropriate. (at [32])

If anything, the focus by Environment Judges on deterrence as the primary purpose of sentencing RMA offenders has increased. For instance, in the 2016 case of *Waikato Regional Council v Rymanda Farms Ltd* [2016] NZDC 15056 the Court stated that "[t]he key sentencing principle required to be reflected in the starting point adopted is that of deterrence" (at [36]). This was a case involving the discharge of dairy farm effluent to land in circumstances where it may enter water, contrary to section 15(1)(b) of the RMA, and the harmful effects of such are well-known. However, deterrence has been said to be equally, if not more, applicable for cases where the harm is more indirect. In the 2017 case of *Tauranga City Council v Jacko Basil Holdings Ltd* [2017] NZDC

3273, where the charges involved the use of a property as two independent dwelling units in breach of the Tauranga City Plan, contrary to section 9(3) of the RMA, the sentencing Judge stated:

> [17] The principles of the Sentencing Act also apply and in particular the key principle of deterrence has been identified in many cases as being the first and most important principle. This must be particularly true in respect of those offences which do not have direct environmental harm but represent wider policy goals for districts such as the density rule.

The primacy of deterrence can also be seen in the 2016/2017 sentencing decisions that were available (details of which are set out below). Of the 39 cases reviewed, where a purpose of sentencing was explicitly mentioned (in 27 of the cases) deterrence was always one of them. Indeed, in 17 of these cases it was the only purpose mentioned.

While judges impose sentences, and therefore have a significant—although as will be argued in Chapter 5 not sole—role to play in the severity of the offences, local authorities have the most influence when it comes to certainty, deciding how to monitor for breaches, when a breach will attract formal enforcement action and what action this will be (as discussed later in this chapter). Accordingly, it is necessary to also consider why local authorities say they are prosecuting, and their enforcement policies reveal that deterrence, rather than desert, is typically emphasised. For instance, the Waikato Regional Council uses a "compliance pyramid" that has been adapted from Ayres and Braithwaite's responsive regulation approach (1992), and which places "Prosecution" at the apex with the goal of "[g]eneral and specific deterrence" (2016, 11). Similarly, the West Coast Regional Council has three factors that need to be considered when deciding whether or not to prosecute, namely the effect on the environment, the degree of culpability of the alleged offender and any other circumstances, which includes not only the public interest and the likelihood the defendant will be discharged without conviction, but also the "need for a deterrent" (2013, 9).

This is not universal. For instance, Environment Canterbury's enforcement policy states that the "fundamental objectives of a criminal prosecution" are "[t]o punish those who deserve punishment for their offences" and "[t]o protect the community and provide justice for victims of offending" (2010, 5), although that council does, earlier in its policy, state that "providing deterrence through appropriate penalties" is a desired outcome of compliance, monitoring and enforcement generally (2010, 1). Other councils do not give a specific purpose for prosecution, although deterrence, and not punishment, typically features throughout their enforcement policy. By way of example, one of the matters Environment Southland considers when deciding on compliance or enforcement action is whether there is "a need for a wider general deterrence required in respect of this activity or industry" (2017, 17).

Perhaps the strongest indication of council priorities comes from the framework developed by the Compliance and Enforcement Special Interest Group (2016). This Group is made up of 14 regional councils and unitary authorities, and together these local authorities account for the vast majority of all RMA prosecutions undertaken—for instance, in 2016/2017 regional councils and unitary authorities brought 65 of the 71 prosecutions initiated (Ministry for the Environment 2018b). The framework the Group has developed emphasises "maximising deterrence" in its "[p]rinciples to guide compliance operations" (2016, 9) and specifically references deterrence (and no other purposes) when setting out the relevant factors, over and above those that generally guide enforcement decision-making, if prosecution is being considered (2016, 14).

More recently, the Ministry for the Environment has published its *Best Practice Guidelines for Compliance, Monitoring and Enforcement under the Resource Management Act 1991* (2018a), which list the "two principal purposes of prosecutions" as being "to punish the offender and to deter potential reoffending by the offender or others" (2018a, 93). On one reading, this is somewhat confused, as it extols local authorities to pursue both desert and deterrence, which rarely coincide. However, it could also be read as these being the two principal purposes of prosecutions *generally*, of which local authorities should pick one in a particular case. But if so, no guidance is given as to whether one or other should be pursued, and the position is not made any clearer by the overall purposes of compliance, monitoring and enforcement set out in the Guidelines, which include to demonstrate "the consequences of non-compliance with the RMA and [provide] both general and specific deterrents, to prevent future offending" (at 12). In any event, there is no indication that local authorities are increasingly favouring desert when they come to update their enforcement policies, despite the mixed approach set out in the Guidelines. For instance, the Marlborough District Council's Enforcement Policy (2018), which was published after the Ministry's Guidelines were issued, is in a similar vein to that published by the Waikato Regional Council, adopting the compliance pyramid with prosecution at the top and with the goal of "[g]eneral and specific deterrence" (at 17).

That deterrence is the primary rationale for sentencing and prosecuting is understandable, given this was the only purpose emphasised in the Explanatory Note to the Resource Management Bill. It is also no surprise in the global context (see, for instance, Lynch et al. 2017, 212; Stretesky et al. 2013, 236; Stretesky 2006; Du Rées 2001). However, there is a rider to this, in that there are likely to be specific cases where other reasons (mainly desert) carry more weight. However, such cases are sufficiently rare (on average, less than one a year) that critiquing the effectiveness of the offences using deterrence theory will not be unduly distorted by them.

Calculating Deterrence

If deterrence is the goal, then the effectiveness of the offences can be gauged by assessing their expected cost, which as noted in Chapter 3 is made up of the quantum of the penalty ("severity") and the probability of having a penalty imposed ("certainty"). Before undertaking this analysis, in the following two sections, it is necessary to look more closely at what makes up the severity and certainty of a penalty, and outline where the data for this and the next section is drawn from.

There are a number of elements involved with "severity". The most obvious is the formal penalty that will be imposed, which in regulatory matters is typically a fine. However, where it is a criminal offence and the offender is an individual, the sentence could be non-monetary.

In the New Zealand context, such sentences are split into "community-based" (community work, supervision, intensive supervision or community detention: see section 44(1) of the Sentencing Act 2002), home detention and imprisonment.

From an economic perspective, monetary penalties and non-monetary penalties are interchangeable, but the former are said to generally be preferable because they are less expensive to impose (Becker 1968, 193). For instance, a fine is a simple transfer of funds from the offender to the State (although the costs of recovering it need to be taken into account), whereas the non-monetary sentences require expenditure on staffing, equipment and infrastructure. However, the interchangeability of monetary and non-monetary penalties is not necessarily a view shared by the public, and the same applies within non-monetary penalties. Indeed, in *R v Conway* [2005] NZRMA 274 (CA) the Court observed that there "is a world of difference in the minds of most members of the community, between a sentence of imprisonment and a sentence of community work" (at [65]).

In more recent times, it has been recognised that severity goes beyond the formal penalty imposed and includes other monetary costs (for instance, legal expenditure in responding to a charge) and non-monetary costs (for instance, the inconvenience factor of being "hassled" by a regulator) (Ogus and Abbot 2002, 290). Further, individuals may want to avoid the shame of being caught breaking the law, and businesses can find there is an impact on profitability from negative publicity surrounding a court case (Macrory 2006, 83). Finally, where a person is subject to a criminal process there may be a stigma that does not necessarily attach to other proceedings (Posner 1985, 1205). The size of these "costs" will depend on a number of factors, such as the type of breach committed and how it is perceived by the offender and the public, and they can be far greater than any fine imposed (Baldwin 2004, 371; Thornton et al. 2005; Shover and Routhe 2005).

The "certainty" aspect of the expected cost also has a number of component parts. As noted when discussing enforcement in Chapter 3, regulators will usually need to choose where to monitor (and accordingly where not to monitor) and how intensively, meaning many breaches will go undetected. Further, when a breach is identified, formal enforcement action does not follow as a matter of course. Finally, even if a regulator seeks to impose a sanction, it may not be proven. Therefore, it is more accurate to consider a number of probabilities that make up "certainty"—that the non-compliance is detected, that enforcement action is taken, and that this enforcement action results in a penalty (Abbot 2009). Dividing the stages of enforcement up in this way also explains why, even if the probability of a penalty being imposed was zero (for instance, if a regulator refused to ever prosecute) there may still be deterrence, as some regulatees will want to avoid the shame of being caught breaking the law even if they do not receive a formal sanction (Braithwaite and Makkai 1991, 12).

The data for the following two sections (severity and certainty) is based primarily on five sources. The first was local authority surveys published by the Ministry for the Environment between 1995 and 2013, which recorded, among other things, matters such as the number of RMA complaints and the number of prosecutions initiated. The second was four studies into the use of prosecutions under the RMA, which were published by the Ministry for the Environment and covered the period from 1 October 1991 to 30 September 2012. These recorded (among other things) the highest, lowest and average fines awarded, both by charge and by prosecution, and also recorded the number of cases where non-monetary and other penalties were imposed. Given the last prosecution study was published in 2013, it was necessary to obtain more recent data, and the third and fourth sources cover the period after the last prosecution study ended (30 September 2012) to 30 June 2017. The third source was information provided by the Ministry of Justice, pursuant to a request under the Official Information Act 1982, on the number of RMA charges filed and resolved and the number of persons/organisations who faced RMA charges during this time period (Ministry of Justice, Letter to author, 23 February 2018). This data was provided in aggregate form for each of the financial years 2012/2013 to 2016/2017. The Ministry also provided data on how many of each type of sentence was imposed, the average length of the non-monetary sentences and the average size of the monetary sentences. The fourth source was RMA sentencing decisions issued during the 1 October 2012 to 30 June 2017 time period, which were obtained from Westlaw, LexisNexis and NZLII databases. The searches that were undertaken identified 224 first instance full-text sentencing decisions and 18 appeals against sentence. From these, the 39 available decisions issued during the 2016/2017 financial year were chosen for specific analysis. The fifth and final source was data the Ministry for the Environment has been collecting since 2014 for a National Monitoring System. This includes similar data to that obtained in the local authority surveys, but is more comprehensive and attributes it to individual local authorities and specific sections of the RMA breached. Data to the end of June 2017 was used.

As White notes "adequate data collection and analysis forms part of what is needed if environmental prosecution and sentencing is to move forward" (2010, 379), and happily the data has improved as time goes on. But it is important to make clear that each of the sources has its limitations. Among other things, date ranges differed, questions changed, interpretations varied by council and recording errors have been made. However, as noted in Chapter 1, there were enough sources to corroborate—or otherwise—any questionable figures that arose. Further, the purpose of assessing the severity and certainty of the offences is not to come up with an exact figure representing the expected cost of the offences. Rather, the point is simply to get an indication of whether the offences are, at a national level, a good or a bad deterrent. Having dealt with any major issues with the data, it was possible to do this.

Severity

We can begin with severity and, as noted in Chapter 2, there are a variety of penalties that can be imposed following a successful RMA prosecution. The first of these is a fine, and while originally the maximum was limited to $200,000, in 2009 this was increased and split between natural and non-natural persons, with the former raised to $300,000 and the latter $600,000. Natural persons are also liable to imprisonment for a term not exceeding 2 years, which triggers the possibility (under the Sentencing Act 2002) of home detention and community-based sentences as alternatives. For both natural and non-natural persons, the Court can also, or instead, make an enforcement order requiring the defendant to take or cease action (as the case may be), or order a consent authority to review the defendant's resource consent (if the offence involves an act or omission that contravenes the consent). A resource consent review can result in the cancellation of the consent if there are significant adverse effects on the environment resulting from the exercise of the consent. The last formal penalty is the conviction, which will be imposed on guilty defendants unless they successfully apply for a discharge without conviction pursuant to section 106 of the Sentencing Act 2002. Finally, as noted in the preceding section, the severity of a sanction includes any informal penalties, again both monetary and non-monetary, felt by the recipient. All of the formal and informal penalties must be taken into account, although as one moves into the latter it becomes harder to assess the impact, given the severity becomes increasingly dependent on the specific circumstances of the offender.

Fines

The fine has been, and remains, by far the most common penalty imposed by the courts for RMA offending. For instance, it would appear from the prosecution studies that, between 1 October 1991 and 30 September 2012, at least 80%, and potentially 90% or more, of successful prosecutions resulted in a fine. Since then, other than in 2013/2014 when fines accounted for 82% of the "sentences recorded", the percentage of fines imposed in successful prosecutions has been greater than 90%. Indeed, in the final year of data considered, 2016/2017, no sentence other than a fine was recorded as having been imposed. This preference for fining defendants is unsurprising, as the same is the case in many other countries, including the United States of America, the United Kingdom, Belgium, Canada and Australia (White 2017), but it would appear that the courts in RMA cases are taking matters to the extreme.

Fines can range from the significant to the minimal. At the former end of the spectrum, the highest fine imposed in the period studied was $300,000 in *Maritime New Zealand v Daina Shipping Company* DC Tauranga CRI-2012-070-1872, 26 October 2012, and this was closely followed by a fine of $288,000 in *Bay of Plenty Regional Council v Mobil Oil New Zealand Ltd* [2016] NZDC 8903. At the other end of the spectrum, fines of as little as $5,000 were still being imposed in 2016/2017, with a number of other fines under $10,000, $15,000 and $20,000 also imposed that year.

Another way to look at fine levels is to consider the average fine per prosecution. This is preferable to considering the average fine per charge, as, in the usual course of events, the fine reflects the offending rather than the number of charges the defendant is convicted of. Average fine per prosecution is also preferable to the average fine per defendant, as some prosecutions have multiple defendants and the courts have held that where the defendants are "one and the same for practical fiscal purposes…the total penalty imposed must be tailored accordingly" (*Kiwi Drilling Company Ltd v R* (1997) 4 ELRNZ 23 (CA) at 28), which typically means dividing the one penalty between the defendants. Put another way, whether one defendant is convicted of one charge or five (related) defendants are convicted of five charges each, if the overall offending is the same then the total penalty imposed in each case should be the same.

Average fines per prosecution is also the approach taken in the prosecution studies, and these show a steady increase since the RMA was enacted. In the first study (1 October 1991 to 30 June 2001) the average fine per prosecution was $6,500, in the second study (1 July 2001 to 30 April 2005) it was $8,167, in the third study (1 May 2005 to 30 June 2008) it was $12,463 and in the fourth study (1 July 2008 to 30 September 2012) it was $21,622. It was during the period covered in the fourth study that the maximum penalty for RMA offending increased from $200,000 to the split penalty of $300,000 for natural persons and $600,000 for non-natural persons. The study notes that during this period, and excluding cases where the Court considered the defendant to be of "poor" financial standing or that the offending was "accidental", the average total fine before the increase was $19,789 per prosecution and after the increase it was $28,792 per prosecution.

There is no data as to the average fine per prosecution since 30 September 2012, but an analysis of the sentencing decisions available for 2016/2017 shows that it has continued to increase. In the 38 (out of

39 available) cases where the primary penalty was a fine, the average fine per prosecution was $36,268. On the most generous interpretation—that is, one natural person defendant—this amounts to 12.1% of the maximum penalty of $300,000, with this percentage steadily decreasing where the prosecution involved multiple defendants (which means the penalty will likely have been divided between them) and/or non-natural defendants (which increases the maximum penalty on each charge to $600,000) and/or continuing offences (whereby the Court can impose an additional penalty of up to $10,000 for every day or part of a day during which the offence continues). While it is true that, in the usual course, "the maximum penalty, being reserved for the worst class of case in the offence category, is of only indirect and sometimes marginal relevance to day-to-day sentencing" (Law Commission 2006, 18), in RMA sentencing (and regulatory offending more generally), where prosecution is typically reserved for the most serious offending, it takes on greater importance. This last point is returned to in the next chapter, but for now it suffices to conclude that, although fine levels are increasing, they are still relatively modest.

Non-monetary and Alternative/Additional Sentences

As noted in the first paragraph of this section, instead of (or in addition to) a fine, natural persons can be sentenced to a non-monetary penalty, such as community work, community detention, home detention or imprisonment. Pursuant to section 10A(2) of the Sentencing Act 2002, such penalties are considered more restrictive than a fine, with only a "discharge or order to come up for sentence if called on" lower in the hierarchy of sentences. When considering these non-monetary penalties, it is important to note that severity includes not only the length of the term of imprisonment or community detention or home detention, or the number of hours of community work, but the frequency with which these penalties are imposed (Von Hirsch et al. 1999, 6).

The most serious sentence a natural person can receive for offending against the RMA is a term of imprisonment. Such a sentence is at the apex of the sentencing hierarchy, being reserved for those cases where it meets the relevant purposes of sentencing in section 7 of the Sentencing Act 2002, those purposes cannot be achieved by a sentence other than imprisonment, and no other sentence would be consistent with the application of the relevant principles in section 8. Account must also be taken of the "desirability of keeping offenders in the community so far as that is practicable and consonant with the safety of the community" (section 16(1)). In some jurisdictions it is not uncommon for a fine to be coupled with imprisonment—for instance, in Belgium between 2003 and 2007 approximately 10% of individuals convicted of environmental offences were sentenced to imprisonment and a fine (Billiet et al. 2014)—but in New Zealand it is typically an alternative.

The prosecution studies do not expressly record the number of natural persons sentenced to imprisonment, but from reviewing these studies it is clear that, between 1 October 1991 and 30 September 2012, this sentence was rarely used. It was not used at all during the first period (1 October 1991 to 30 June 2001), where there were 375 prosecutions, and was used only twice during the second period (1 July 2001 to 30 April 2005), where there were 171 prosecutions, the first time for three months and the second time for 20 weeks (reduced on appeal to 12 weeks). Imprisonment was also used just twice during the third period (1 May 2005 to 30 June 2008), where there were 260 prosecutions, for six weeks and eight weeks, and in both instances the defendant was already serving a term of imprisonment, which essentially ruled out alternative options. Indeed, the sentencing Judge in the first of these cases, *R v Feeney* DC Hamilton CRI 2006-019-6265, 7 April 2008, noted that the defendant's offending was of a sort that would not normally result in a term of imprisonment, and the same would seem to be true in the second case (*Thames-Coromandel District Council v May* DC Auckland CRI-2008-004-7940, 5 June 2008), as the offending only involved having more than the permitted number of buildings on a property. Finally, imprisonment was used once during the fourth period (1 July 2008 to 30 September 2012), where there were 429 prosecutions, for a term of six and a half months, the defendant being one of the people who had been imprisoned previously.

It would appear that no further terms of imprisonment had been imposed by the end of the 2016/2017 financial year. As such, putting the two defendants that were already incarcerated when they were sentenced on RMA offending to one side, in the first 26 years of the RMA only two defendants had been sentenced to imprisonment (one of these twice), for terms ranging from ten weeks to six and a half months. Such terms are between just under 10% and just over 25% of the maximum penalty of two years.

Despite section 339(1) of the RMA referring only to imprisonment as an alternative to a fine for natural persons, section 339(4) and the provisions of the Sentencing Act 2002 allow the Court to instead impose a sentence of community work, a different community-based sentence or home detention. As with fines and imprisonment, there is guidance provided in the Sentencing Act 2002 as to when it is appropriate to use these sentences. These sentences fall between imprisonment and fines in terms of restrictiveness, with home detention being a more restrictive sentence than intensive supervision and community detention, which in turn are more restrictive sentences than community work and supervision.

Such sentences have been used more than imprisonment, but still relatively rarely. In the period covered by the first prosecution study (when the Criminal Justice Act 1985 was the primary sentencing statute), there were eight sentences of community service and three sentences of periodic detention imposed. In the second study period (by which time the Sentencing Act 2002 had come into force) there were four sentences of community work imposed, and the third study records that there were 12 sentences of community work handed down. Finally, during the fourth period there were 22 sentences of community work imposed (one of which was substituted for a fine after a failure to complete the work), two of which also included sentences of community detention.

Turning to the most recent data from the Ministry of Justice, which records the most serious sentence a defendant received, 391 offenders (which includes both natural and non-natural persons) were sentenced between 2012/2013 and 2015/2016, with two receiving home detention, six receiving community detention and 11 community work. In some of these cases the defendants will have also received a fine, but for the most part a community-based sentence is only imposed because the offender does not have sufficient funds to pay a monetary penalty (see, for example, *Southland Regional Council v Egginton* [2015] NZDC 14393). In these instances, the Court has noted that it is not possible to directly compare the number of hours of community work imposed with the fines that would otherwise have been imposed, and the same reasoning would apply to months of community detention or home detention. The average number of hours of community work imposed per offender during this period was 152 and the average days of home detention and community detention were 51 and 100 (respectively). No offenders were sentenced to community-based sentences or home detention in 2016/2017.

For both natural and non-natural persons, there are two other penalties that can be imposed, either additionally or alternatively, following a successful prosecution. One is an enforcement order under section 314 of the RMA and the other is an order requiring a consent authority to serve notice, under section 128(2) of the Act, of the review of a resource consent held by the person, but only if the offence involves an act or omission that contravenes the consent. Looking first at enforcement orders, as noted in Chapter 2 there are a number of different things these can be used for, such as requiring a person to cease doing, or prohibiting them from commencing, something that is contravening (or is likely to contravene) the Act, a rule or a resource consent. Alternatively, the enforcement order could require a person to do something to ensure he or she complies with the Act, a rule or a resource consent. Further options include that the person could be required to remedy or mitigate any adverse effect on the environment they have caused, or pay the actual and reasonable costs and expenses of remedying or mitigating any adverse effect on the environment caused by that person's breach.

Enforcement orders on sentencing are reasonably uncommon. For instance, in the case of *Bay of Plenty Regional Council v Whitikau Holdings Ltd* [2018] NZDC 3850 the sentencing Judge observed that for "virtually all RMA offending, the fact that reparation and enforcement orders are available is a theoretical, but not practical, reality" (at [236]). This is borne out somewhat by the prosecution studies, which note that there were 36 enforcement orders as part of sentencing in the first period (out of 375 prosecutions), 21 in the second period (out of 171 prosecutions), 38 in the third period (out of 260 prosecutions) and 32 in the fourth period (out of 429 prosecutions). No data has been kept on the number of enforcement orders made on sentencing since then, but it can be observed that eight enforcement orders were imposed in the 39 decisions analysed from the 2016/2017 financial year.

When enforcement orders are made, they are typically "compliance-focussed", requiring the offender either to remedy the damage caused or to fix the issue that led to the non-compliance. An example of the former can be found in the case of *Queenstown Lakes District Council v Allenby Farms Ltd* [2017] NZDC 3251 where the Court, in addition to imposing a fine, made an enforcement order requiring the defendant to, among other things, replant various areas that it had unlawfully cleared on its land. An example of the latter is the approach now commonly seen in certain regions whereby the local authority seeks, as a matter of course, an enforcement order in a dairy farm effluent discharge sentencing that requires the defendant to upgrade its effluent management system in accordance with a plan developed by an accredited effluent storage designer.

It is much rarer to see "punitive" enforcement orders being imposed on defendants. While these do not obviously seem to fit within the terms of section 314 of the Act, it is apparent that they are available. For instance, in *Machinery Movers Ltd v Auckland Regional Council* the High Court observed that a sentencing Court has the jurisdiction under section 314 to impose similar orders as those made in the Canadian case of *R v Bata Industries Ltd* (1992) 9 OR (3d) 329, 7 CELR (NS) 293 (Ontario Court (Provincial Division)), where the defendant was required to:

> ...have published on the front page of the Bata newsletter for international distribution the facts leading to this conviction of the company and the named directors, Mr Weston and Mr Marchant, the details of the penalties and the terms of probation;
>
> ...

However, despite such "naming and shaming" being generally considered to be a deterrent (see *Environment Protection Authority v Dyno Nobel Asia Pacific Pty Ltd* [2017] NSWLEC 64 at [163]), similar orders have since been made under the RMA only occasionally. There are only two instances present in the legal databases—*Auckland Regional Council v Nuplex Industries Limited* DC Auckland CRN2004066321, 18 March 2003 and *Auckland Regional Council v URS New Zealand Limited* DC Auckland CRI-2008-004-13603, 23 July 2010 and 6 August 2010—and the order in the latter case was quashed when the defendants' appeals against their convictions were successful (*URS New Zealand Limited v Auckland Regional Council* [2012] NZHC 723).

A different sort of punitive enforcement order was issued in *Bay of Plenty Regional Council v Hood* [2017] NZDC 25156. In this case, the defendant was ordered to cease pig farming, remove all pigs from the property and after that not keep any pigs on the property (with no end date specified). While it can be observed that the defendant agreed to the making of such an order, as he intended to stop pig farming anyway, as the Judge observed "[i]t is important that the sentence be appropriate in terms of the Sentencing Act 2002 and the Resource Management Act 1991 rather than simply the result of an agreement between the parties" and that the terms (subject to one minor amendment) were "appropriate" (at [27]). However, no further enforcement orders of this nature appear to have ever been made.

The enforcement order in *Hood* was, perhaps, also easier to make in that case because the defendant was not operating pursuant to a resource consent. However, where the offending results from the breach of a consent there is the sentencing option of an order requiring a consent authority to serve notice, under section 128(2) of the Act, of the review of a resource consent held by the person. This can fairly be described as "punitive" because section 128(2), in combination with section 132(4), allows the consent authority, following the review, to cancel the resource consent if "there are significant adverse effects on the environment resulting from the exercise of the consent". This is, therefore, arguably the most serious penalty a court can impose (along with a Hood-type punitive enforcement order), and the purpose of it, when it was introduced in 2009, was to provide another deterrent option (Cabinet Paper 2009, 25). However, a search of the relevant legal databases has revealed only one record of a local authority asking for the Court to order such a review, in the case of *Taranaki Regional Council v A J Cowley Ltd* DC New Plymouth CRI-2011-043-2428 & 3192,

15 December 2011, and it was declined because there was no "obvious connection between the offending or circumstances surrounding it and the conditions of consent" and "the consent was due to expire in 2014 in any event" (at [29]).

It may be that rather than reviewing, and potentially cancelling, the resource consent, the conviction is instead being considered by the local authority when it comes time for the consent-holder to renew its (or apply for a new) consent. Accordingly, this is an appropriate time to consider the severity of an RMA conviction in and of itself, but the conclusion regarding non-monetary penalties is two-fold. First, non-monetary sentences for individuals, whether they be imprisonment, home detention, community detention or community work, are infrequent, and becoming even rarer, meaning that their actual length is largely irrelevant (although often towards the lower end). Second, where an enforcement order is made, it is likely to be compliance-focussed, rather than punitive, and a resource consent review, which can also be punitive, does not appear to have ever been ordered. As such, none of these penalties are contributing much to the overall severity of the RMA's offences.

Convictions and Informal Penalties

The conviction is the last formal penalty, although in some ways it could have been discussed first, as section 11(1) of the Sentencing Act 2002 provides that the initial decision a sentencing court must make is whether or not to convict the defendant. However, it is appropriate to deal with it last as it is the formal penalty that appears to carry, in the usual course, the least punitive effect. This may seem at odds with the fact that defendants often apply for a discharge without conviction, which suggests that at least these offenders believe an RMA conviction to be a significant penalty and one they wish to avoid, but this can be explained in another way. This is that a discharge without conviction means that there will not be any fine or non-monetary sentence imposed (although costs and/or reparation can be ordered), and so it may be that some defendants apply for a discharge without conviction to avoid receiving a "primary" penalty.

Indeed, even if the defendant is unsuccessful, there may be a tactical benefit in making such an application as it implies that the offending is low level, on the basis that a discharge without conviction can only be granted if "the court is satisfied that the direct and indirect consequences of a conviction would be out of all proportion to the gravity of the offence" (Sentencing Act 2002, section 107), and could have the effect of orientating the sentencing submissions accordingly.

In one sense, it is understandable that defendants are most concerned about the "primary" penalty they face (typically, as noted above, the fine), as the effects of it will be immediate compared with those of a conviction. But a conviction is typically considered to be a penalty, quite apart from any other sentence imposed, because it labels the defendant, to the wider community, as a "criminal" (Simester et al. 2013, 5), and for many offences this has real and serious effects. For instance, a conviction can have formal consequences, including travel and employment/contracting implications, it might affect the defendant's ability to obtain insurance (and the premiums that must be paid), and it can restrict one's ability to be a member of professional organisations. Further, a conviction may have informal consequences, such as arousing in the offender a sense of shame or embarrassment, and it can result in adverse treatment by members of the offender's community.

However, these consequences appear to be of less importance in RMA offending. Starting with the formal consequences, these do not typically appear to be significant. For instance, the sentencing Judge in *Tauranga City Council v Jacko Basil Holdings Ltd* [2016] NZDC 17674 expressed uncertainty as to whether an RMA conviction will impact on a defendant's ability to travel, and other restrictions, such as relating to employment and tender applications, will likely need to be assessed on a case-by-case basis. But where one would expect an RMA conviction to have the greatest impact, namely on future resource consent applications, it is not formally recognised. Earlier in this section it was noted that a conviction appears to never lead to a consent review, despite there being statutory provision for this, and nor are RMA convictions explicitly a relevant factor for a consent authority to take into account in considering an application for a resource consent under section 104(1) of the Act. A conviction (and non-compliance generally) may be taken into account as "any other matter the consent authority considers relevant and necessary to determine the application", but this will depend on the consent authority's disclosure requirements and knowledge of the applicant, and the emphasis the consenting arm of the council places on previous history (as opposed to the economic benefits that the application will bring). A recent survey found that only 59% of councils took into account the compliance history of a resource consent applicant, and in some instances this "taking into account" appears to only relate to setting the conditions of, not whether to grant or decline, the consent (Brown 2017, 51; see also Ministry for the Environment 2016a, 39).

In terms of informal consequences, it is no doubt the case that some people convicted of RMA offending will feel ashamed or embarrassed, and that some businesses will find their reputation adversely affected. However, it is questionable whether the shame, embarrassment or reputational damage is because the person/business has received a conviction, as opposed to the mere fact they have had enforcement action taken against them (and been penalised). For instance, such consequences would typically arise following media coverage of a prosecution, but the focus of reporting is often on the nature of the offending and the penalty imposed, rather than the fact the offender received a conviction. Indeed, given the regulatory nature of the RMA, coupled with the fact that the usual penalty imposed is a monetary impost, the public may not even be aware that these are even criminal offences. It is, accordingly, convenient to discuss shame, embarrassment and reputation damage more broadly in the context of "informal penalties".

The first informal penalty is the direct costs involved in defending RMA charges, and these can be considerable. By way of example, in *Wallace Corporation Ltd v Waikato Regional Council* [2012] NZHC 1420 it was noted that the defendants incurred legal costs of $426,757 and expert costs of $146,240 in unsuccessfully defending an (albeit difficult and complex) prosecution, and a further $140,897 in legal costs in successfully appealing against the first instance decision. One would expect, however, that in the usual course these costs would typically be less than those incurred by the prosecutor, given the local authority must be the one to drive the case (for instance, gather sufficient evidence, disclose the relevant material to the defendant, prove the charges beyond reasonable doubt, and so on). Further, the costs incurred will not necessarily be proportionate to the seriousness of the charge—a major contaminant discharge may only need the entry of a guilty plea and a relatively simple plea in mitigation, whereas a technical District Plan breach may need a lengthy defended hearing with expert evidence (and then a plea in mitigation if found guilty). Rather, the costs incurred will more likely be a function of the importance of the case to the defendant and resources available to it.

Similarly, how much of an effect "hassle" costs have will typically be related to the type of defendant. Large firms will be accustomed to dealing with regulators and the court process, and may have experts in their staff for this purpose (who often will have a local government background), meaning the business of the firm can go on relatively uninterrupted. On the other hand, individuals and small-to-medium sized enterprises may find such interactions to be unfamiliar and uncomfortable, and may need to assign such responsibilities to frontline staff, diverting them away from their primary tasks. The converse is true for reputational damage. One would expect that individuals and small-to-medium sized businesses are more likely to be concerned about the primary penalty that may be imposed, while large firms—especially those that consider themselves good corporate citizens—may find any reputational damage that results from enforcement action to be the biggest penalty. This may explain why the defendant in *Bay of Plenty Regional Council v Mobil Oil New Zealand Ltd* never appealed against the fine of $288,000, which would risk further publicity, despite the sentencing Judge adopting a starting point for the offending well in excess of what even the prosecutor had sought. It is, of course, more complex than this, as some small family businesses with a long history in a particular community will place great weight on their reputation (Whyte 2010, 186), and some large corporations will consider that all the public really cares about is the price and quality of the product they manufacture, not how they do so (Baldwin 2004, 364).

Whatever the case, for reputational damage to be brought about the public, or at least the regulated community, needs to be made aware of any enforcement action, and yet publicity is sporadic. The starting point is that there is no centralised reporting of RMA prosecution outcomes, with the prosecution studies apparently having been discontinued (Brown 2017, 53) and the Ministry for the Environment only publishing, by way of the National Monitoring System (and before that the Local Authority Surveys), data on the number of prosecutions brought, not the results of these cases. Nor do the courts publish any analysis of the outcomes achieved, with the 2017 Environment Judges' Annual Review devoting only two paragraphs in the 32-page report to prosecutions (Environment Court of New Zealand 2017, 10), and the 2017 Registrar of the Environment Court's Annual Report mentioning these in only one sentence in a 12-page report (Registrar of the Environment Court 2017, 6). This is presumably because the prosecutions are heard in the District Court, but RMA cases are given no specific attention in the reports issued by that Court either.

The task is therefore left up to the local authorities to provide details of the outcomes of the cases they have brought, but only a small minority of councils publish regular enforcement reports detailing their successful prosecutions. Given it is unusual to find members of the public in attendance at an RMA sentencing hearing, the community is, accordingly, left to rely on press releases and media reports for individual cases. While these appear to be issued more frequently than in the past, they are not comprehensive and, as with the nature of the news cycle, will become "lost" if they are not seen in the short period after they are published.

Finally, if a member of the public wanted to read a sentencing decision after seeing a summary in an enforcement report, press release or media report, there is no guarantee they would be able to do so. Local authorities usually do not publish sentencing decisions on their websites (but see, as an exception, the Bay of Plenty Regional Council) and only a limited number of RMA sentencing decisions are available on public websites. Further, these cases are listed with all other District Court sentencing decisions, meaning that the searcher needs some level of knowledge to be able to find them.

The picture painted here about informal penalties is one of variability. Some defendants will spend a lot of money defending the charges and feel deeply embarrassed for having offended, while other defendants will not care that they have been prosecuted and they will not suffer any noticeable impact on their business. To an extent, this variability is to be expected. However, it is suggested that matters are not helped by the inconsistency between local authorities in terms of how successful prosecutions are promoted, and the lack of a centralised system for reporting on outcomes.

This brings us full circle. The foregoing has shown that the severity of the RMA's offences primarily comes from the fine that is imposed. Yet while there have been gradual increases to average fines levels, these could still only be described as modest. And this deterrent value depends on prospective offenders being aware of these consequences, which is patchy at best, and so the severity of the RMA's offences appears to be no more than moderate.

Certainty

As set out earlier in this chapter, there are three main factors that make up the certainty of having a criminal sanction imposed—the probability that a breach is detected, the probability that it results in formal enforcement action and the probability that this enforcement action results in a criminal penalty being imposed. And as with severity, each of these must be taken into account.

Detection of Breaches

Starting with the detection of breaches of the RMA, local authorities generally rely on a combination of methods, some of which are "reactive" and others which are "proactive" (Ministry for the Environment 2018a). In terms of the former, people can make complaints about suspected breaches to the relevant local authority, with many councils operating "hotlines" 24 hours a day, 7 days a week to receive these. Such complaints will typically come from a member of the public, given there is often no direct victim of RMA offending, and could relate to (suspected) breaches of resource consent conditions or plan rules, although they are typically more general than this (for instance, where someone notices that a waterway is discoloured). Councils are expected to investigate such complaints and are under a duty to keep available at their offices not only a summary of all written complaints received during the preceding five years, but also information on how they dealt with each complaint (RMA, section 35(5)(i)). The Ministry for the Environment has endeavoured, at least in the National Monitoring System, to get local authorities to separate complaints out from monitoring when reporting—that is, if a complaint triggers monitoring, then it should be recorded as a complaint, not monitoring—and therefore there are reasonable records relating to the numbers of complaints received. Since 2010, these have ranged from 24,269 per year (in 2010/2011) to 31,446 per year (in 2015/2016), with there being 27,857 complaints in 2016/2017, the last year of data considered in this book.

This is a large number of complaints, and it gives the impression that local authorities are detecting a lot of RMA non-compliance through this means. However, a complaint is not the same as a detected breach, and not all of these complaints are followed up in detail, or even at all. As the Ministry for the Environment has noted:

> The majority of councils have adopted a risk-based approach to complaints response … and respond to complaints depending on their risk and priority. Most councils stated that they respond to most complaints within two to three days. Some councils reported, however, that due to a lack of resources and staff capacity, there is a backlog of complaints that had not been investigated. (Ministry for the Environment 2016a, 20)

This means that there will be breaches of the RMA that, despite being notified to the local authority (in the form of a complaint), are not "detected". Indeed, even if the breach is established, should the party responsible for it not be located then it will still not reach the threshold for detection.

Part of the reason why not all complaints can be followed up is that local authorities spend a lot of their time monitoring resource consents. The Ministry for the Environment has reported on the monitoring of resource consents from the start of the local authority surveys, but the details of what was reported changed so frequently that the numbers only became meaningful since the introduction of the National Monitoring System. Focussing then on this data, the latest figures considered in this case study record that, during the year ended 30 June 2017, there were 99,986 resource consents nationwide that required monitoring.

Of these 99,986 resource consents that required monitoring during that year, 77,013 were in fact monitored. It can immediately be seen, therefore, that a little over a fifth of consents (approximately 22,000) that should have been monitored were not. However, it is likely that the number is greater than this, as it is apparent that some councils monitored individual resource consents more than once, yet this is only reflected in the monitored (as opposed to the required monitoring) column of the National Monitoring System data. For instance, in 2016/2017 Wellington City Council had 343 resource consents that required monitoring, yet 490 resource consents were in fact monitored.

From the 77,103 resource consents that were monitored in 2016/2017, 13,289 were found to be in "non-compliance", with this defined as being "in non-compliance with at least one condition" (Ministry for the Environment 2016b, 46). This would appear to suggest that resource consent monitoring detects a considerable number of RMA breaches. However, the first point to note is that a binary system of compliance or non-compliance does not give any indication as to the level of non-compliance. Self-evidently, significant non-compliance is worse than moderate non-compliance, which is worse than low risk non-compliance, and it is not clear how many of these 13,289 breaches fell into which category. More importantly for present purposes, though, a distinction needs to be made between (what could be termed) "technical" breaches of the RMA and "substantive" breaches of the RMA. The former are those where a consent-holder is not complying with the conditions of its consent and so they are no longer allowed to undertake the authorised activity, and so are "technically" breaching the RMA every time they do. An example of this would be a resource consent that authorised the taking of up to 100 litres of water per day, but only on the condition that records were provided to council of the amount of water taken. A failure to provide the records to the local authority means the consent-holder is in breach of the conditions of the consent, and so any taking of water is a (technical) breach of the RMA (until the breach is rectified). On the other hand, substantive breaches are those where the terms of the authorised activity itself are breached. Using the same example, this would include where the consent-holder took more than 100 litres per day, which would amount to a (substantive) breach of the RMA.

Resource consent monitoring is important as it will help to identify technical breaches, which may well mean that there is a systemic issue that could lead to a substantive breach. However, it will generally not identify substantive breaches unless these happen to occur on the day that the inspection takes place. This is even less likely for those regions where councils warn consent holders, either generally (that is, they could be inspected any time in the following days or months) or specifically (that is, they will be inspected on a certain day), of an impending inspection (as discussed in Ministry for the Environment 2016a, 18). If such a breach occurs on any other day of the year when monitoring is not taking place, it will go undetected by the local authority unless the council is alerted to it by the public (that is, by making a complaint) or by the consent-holder itself. However, there is obviously no guarantee that a substantive breach will be observed by a member of the public, that it will be identified as a breach if seen, and that it will be reported to a local authority if identified. Further, while some consents require the holders to report any breaches, this is not universal and there is no guarantee that reporting will in fact happen.

The final way of detecting non-compliance is by way of general monitoring of plan rules, but this is typically given a lower priority than complaints and resource consent inspections. Such monitoring involves local authorities checking whether people are undertaking: a permitted activity (and, if so, whether they are complying with any required conditions); an activity for which they need a resource consent (but do not have one); or a prohibited activity. This is one of the most difficult tasks for an enforcement officer, as there are often no notification requirements for those undertaking a permitted activity and, self-evidently, those undertaking activities that need a resource consent or activities that are prohibited will not notify the council. Further, the size of many regions makes such general monitoring impractical and, even if such monitoring is undertaken, it can be difficult to determine if a breach has occurred, because permitted activity rules must be written broadly given their general application. Using 2016/2017 as an example, the National Monitoring System data shows that there were only 2,993 instances of general plan rule monitoring undertaken nationwide, and this was by just 14 local authorities (with three of these councils doing so on five or fewer occasions). Indeed, one local authority, Hamilton City Council, was responsible for over half the total instances of general plan rule monitoring (with 1,722). While there seem to be gaps in the data—for instance, no instances of general plan rule monitoring are recorded against the local authority that had the most such instances in 2015/2016, Waikato Regional Council—it seems reasonable to conclude that there is not any significant amount of general plan rule monitoring being undertaken.

Overall, therefore, it would appear that many RMA breaches are not being detected. Complaints are more likely to identify substantive breaches, but local authorities are being overwhelmed by the number they need to respond to. Resource consent inspections do a reasonable job of identifying technical breaches, but will only detect substantive breaches if these happen—coincidentally—on the day of the inspection. Then there is likely to be a great mass of technical and substantive breaches that go unnoticed, as there is little time, and it is somewhat impractical, for local authorities to undertake any significant level of general plan monitoring.

Formal Enforcement Action

However, it is obviously the case that some RMA breaches are detected, and the next step is to determine the probability that a detected breach results in formal enforcement action (as opposed to informal enforcement action or no action at all). It is also important to be clear about terminology here, as some local authorities record written warnings as formal enforcement action (see, for instance, Waikato Regional Council 2017, 64). This is not the approach taken in this book, though, as there is no statutory basis for the giving of a written warning. Rather, for the purposes of this discussion, formal enforcement action includes, on the "directive" route, the options of the lower level abatement notice (issued by an enforcement officer) and the higher level application for an enforcement order (made to the Environment Court), and, on the "punitive" route, the options of the lower level infringement notice (issued by an enforcement officer) and the higher level prosecution (heard in the District Court by an Environment Judge), and a local authority can choose to go down one or both routes (either simultaneously or one after another).

A preliminary problem arises, in determining the probability that a detected breach results in formal enforcement action, in that the precise number of detected breaches is unknown. While the number of complaints is recorded, there is no information, in either the local authority surveys or the National Monitoring System, as to what percentage of these were found to be RMA breaches. Similarly, there is no record of how many of the instances of general plan rule monitoring detected breaches. The only data relating to detected breaches is the number of instances of resource consent non-compliance, which were presumably all technical RMA breaches and, perhaps, some substantive breaches (if, by chance, these occurred on the day the local authority was inspecting the resource consent).

However, in terms of the (unknown number of) detected RMA breaches, it is apparent that there is a considerable amount of informal (or no) enforcement action taken. This must be the case when, by way of example, in 2016/2017 there were 13,289 instances of resource consent non-compliance (and presumably some of the 27,857 complaints and 2,993 instances of general plan rule monitoring resulted in detected RMA breaches) and only 3,344 formal enforcement actions. Further, in some instances multiple enforcement actions will have been taken for one breach (for instance, both an infringement notice and an abatement notice may have been issued, and there may have been a number of each), reducing the number of detected breaches that were formally responded to.

Turning to the make-up of the formal enforcement action, in 2016/2017 the directive route contained 1,876 abatement notices issued and 25 applications for an enforcement order, and on the punitive route there were 1,372 infringement notices issued and 71 prosecutions initiated. It is important to be clear that the number of applications for enforcement orders and the number of prosecutions initiated cannot be directly compared to the number of abatement notices served and infringement notices issued (respectively). This is because, as set out above, one incident may have been dealt with by a number of infringement notices and abatement notices, which will have been counted separately, or by a prosecution or enforcement order application involving a number of defendants and charges, which will have been grouped and counted together. However, even taking this into account, prosecutions (and applications for enforcement orders) are far more infrequent than the lower level types of formal enforcement response.

(Successful) Prosecution

The number of prosecutions has increased since the RMA was enacted. The prosecution studies record that between 1 October 1991 and 30 June 2001 there were 39 prosecutions per year, between 1 July 2001 and 30 April 2005 there were 45 prosecutions per year, between 1 May 2005 and 30 June 2008 there were 82 prosecutions per year and between 1 July 2008 and 30 September 2012 there were 101 prosecutions per year (Ministry for the Environment 2013, 12). The prosecutions studies have (acknowledged) limitations, relying on data sourced from legal databases and information provided by the local authorities, but the local authority surveys (which also have limitations) corroborate this somewhat, recording the number of prosecutions initiated as 81 in 2001/2002, 39 in 2003/2004, 86 in 2005/2006, 167 in 2007/2008, 135 in 2010/2011 and 104 in 2012/2013 (an average of 102 per year over this time period).

However, it would seem that there has been a decrease in the number of prosecutions in recent years. The National Monitoring System records that there were 81 prosecutions initiated in 2014/2015, 57 in 2015/2016 and 71 in 2016/2017. This data also has limitations, but the drop-off appears to be corroborated somewhat by the number of resolutions recorded in the Ministry of Justice data—there were 527 and 643 charges resolved in 2012/2013 and 2013/2014, respectively, which then reduced to 442, 353 and 387 charges in the following three years. Similarly, the number of offenders who had RMA charges resolved against them was 175 and 174 in 2012/2013 and 2013/2014 (respectively), which then reduced to 153, 118 and 145 in the following three years.

Although prosecution is infrequent, when a local authority does bring a case it will usually be successful. The position in the prosecution studies is somewhat unclear because, other than the fact that they do not include every decision, the successful cases and unsuccessful cases together do not add up to the total number of prosecutions, and nor is there any data on prosecutions that were withdrawn by the local authority. However, even taking a conservative approach, over 90% of the prosecutions appear to have been successful. The success rate since 30 September 2012 is unknown, as the Ministry of Justice data includes not only acquittals and charges dismissed in the group "Not proved", but also charges withdrawn by the local authority. There is, though, nothing to suggest that local authorities are achieving fewer guilty verdicts than previously.

However, this high success rate does little to bolster the certainty that a criminal penalty will be imposed when so few prosecutions are commenced. Using 2016/2017 as an example, 71 prosecutions out of 3,344 formal enforcement actions out of an unknown number (but likely to be in the tens of thousands) of detected breaches suggest that the prospect of facing prosecution for non-compliance is slim. Nor is this helped by private prosecutions, which would be one way to augment the enforcement action undertaken by the local authorities. These are the only punitive option available to individuals (or organisations), as the ability to issue an infringement notice is restricted to enforcement officers, and there do not appear to have been any successful private prosecutions in the history of the RMA. As such, when one considers that many breaches are also missed, the certainty that an RMA breach will result in a criminal penalty appears to be low.

Conclusion

It has been shown that the courts and local authorities generally consider that the purpose of the RMA's offences is to deter would-be offenders, which means that the effectiveness of the offences can be assessed by gauging their expected cost. If this expected cost is significant then the offences will be effective, as it will need a correspondingly high expected benefit from breaching the RMA for a rational regulatee to offend. On the other hand, if the expected cost is minimal then the offences will be ineffective, as a correspondingly low expected benefit is all that will be needed before a rational regulatee will breach the Act.

The expected cost of offences is made up of the severity of the penalties imposed and the certainty of imposition, and it has been argued that, for the RMA, the former is no more than moderate and the latter is low. As such, the offences appear to have a minimal expected cost and so are ineffective as a deterrent. Indeed, what is of further concern is that it would appear that the position is not improving. Even though fine levels have been (slowly) increasing, the number of prosecutions, while once rising, has started to decrease again, with such change coinciding (approximately) with the increase in the maximum penalty. This is particularly problematic because, as discussed in Chapter 3, certainty is generally considered to be the more important of the two expected cost variables.

It is necessary to investigate why the expected cost of the offences is minimal, especially given the rhetoric from the courts and local authorities about deterrence being the goal. This is the task of the next chapter, which looks at the reasons why there is a low certainty that a criminal penalty will be imposed and why the penalties are of no more than moderate severity.

References

Abbot, Carolyn. 2009. *Enforcing Pollution Control Regulation*. Portland: Hart Publishing.

Ayres, Ian, and John Braithwaite. 1992. *Responsive Regulation: Transcending the Deregulation Debate*. New York: Oxford University Press.

Baldwin, Robert. 2004. The New Punitive Regulation. *Modern Law Review* 67 (3): 351–383.

Becker, Gary. 1968. Crime and Punishment: An Economic Approach. *Journal of Political Economy* 76 (2): 169–217.

Billiet, Carole, Thomas Blondiau, and Sandra Rousseau. 2014. Punishing Environmental Crimes: An Empirical Study from Lower Courts to the Court of Appeal. *Regulation & Governance* 8 (4): 472–496. https://doi.org/10.1111/rego.12044.

Braithwaite, John, and Toni Makkai. 1991. Testing an Expected Utility Model of Corporate Deterrence. *Law & Society Review* 25 (1): 7–40.

Brown, Marie. 2017. *Last Line of Defence: Compliance, Monitoring and Enforcement of New Zealand's Environmental Law*. Auckland: Environmental Defence Society.

Cabinet Paper. 2009. Reform of the Resource Management Act 1991: Phase One Proposals.

Compliance and Enforcement Special Interest Group. 2016. Regional Sector Strategic Compliance Framework 2016–2018.

Du Rées, Helena. 2001. Can Criminal Law Protect the Environment? *Journal of Scandinavian studies in Criminology and Crime Prevention* 2 (2): 109–126. https://doi.org/10.1080/140438501753737606.

Environment Canterbury. 2010. Compliance Monitoring and Enforcement Guidelines 2010.

Environment Court of New Zealand. 2017. Annual Review Calendar Year 2017.

Environment Southland. 2017. Compliance Policies.

Law Commission. 2006. Sentencing Guidelines and Parole Reform: NZLC R94.

Lynch, Michael J., Michael A. Long, Paul B. Stretesky, and Kimberly L. Barrett. 2017. *Green Criminology: Crime, Justice, and the Environment*. Oakland: University of California Press.

Macrory, Richard. 2006. Regulatory Justice: Making Sanctions Effective.

Marlborough District Council. 2018. Enforcement Policy.

Ministry for the Environment. 2013. A Study into the Use of Prosecutions Under the Resource Management Act 1991: 1 July 2008–30 September 2012.

Ministry for the Environment. 2016a. Compliance, Monitoring and Enforcement by Local Authorities Under the Resource Management Act 1991.

Ministry for the Environment. 2016b. National Monitoring System—Information Requirements: Guidance for the 2016/17 Financial Year.

Ministry for the Environment. 2018a. Best Practice Guidelines for Compliance, Monitoring and Enforcement Under the Resource Management Act 1991.

Ministry for the Environment. 2018b. National Monitoring System for 2016/17.

Ogus, Anthony, and Carolyn Abbot. 2002. Sanctions for Pollution: Do We Have the Right Regime? *Journal of Environmental Law* 14 (3): 283–298.

Posner, Richard. 1985. An Economic Theory of the Criminal Law. *Columbia Law Review* 85 (6): 1193–1231.

Registrar of the Environment Court. 2017. Report of the Registrar of the Environment Court for the 12 Months Ended 30 June 2017.

Shover, Neal, and Aaron Routhe. 2005. Environmental Crime. *Crime and Justice* 32: 321–371.

Simester, A.P., J.R. Spencer, G.R. Sullivan, and G.J. Virgo. 2013. *Simester and Sullivan's Criminal Law: Theory and Doctrine*, 5th ed. Oxford: Hart Publishing.

Stretesky, Paul B. 2006. Corporate Self-Policing and the Environment. *Criminology (Beverly Hills)* 44 (3): 671–708. https://doi.org/10.1111/j.1745-9125.2006.00060.x.

Stretesky, Paul B., Michael A. Long, and Michael J. Lynch. 2013. Does Environmental Enforcement Slow the Treadmill of Production? The Relationship Between Large Monetary Penalties, Ecological Disorganization and Toxic Releases Within Offending Corporations. *Journal of Crime & Justice* 36 (2): 233–247. https://doi.org/10.1080/0735648X.2012.752254.

Thornton, Dorothy, Neil Gunningham, and Robert Kagan. 2005. General Deterrence and Corporate Environmental Behavior. *Law & Policy* 27 (2): 262–288. https://doi.org/10.1111/j.1467-9930.2005.00200.x.

Von Hirsch, Andreas, A. Bottoms, E. Burney, and P.-O. Wikstrom. 1999. *Criminal Deterrence and Sentence Severity: An Analysis of Recent Research*. Oxford: Hart Publishing.

Waikato Regional Council. 2016. Enforcement Policy.

Waikato Regional Council. 2017. 2016/17 Annual Report.

West Coast Regional Council. 2013. Enforcement Policy.

White, Rob. 2010. Prosecution and Sentencing in Relation to Environmental Crime: Recent Socio-Legal Developments. *Crime, Law and Social Change* 53: 365–381.

White, Rob. 2017. Reparative Justice, Environmental Crime and Penalties for the Powerful. *Crime, Law and Social Change* 67 (2): 117–132. https://doi.org/10.1007/s10611-016-9635-5.

Whyte, David. 2010. An Intoxicated Politics of Regulation. In *Regulation and Criminal Justice*, ed. Hannah Quirk, Toby Seddon, and Graham Smith, 162–191. Cambridge, UK: Cambridge University Press.

5

Explaining the Lack of Effectiveness: Constraints and Choices

The conclusion from the previous chapter was that the RMA's offences are ineffective. The primary purpose of the offences, according to the local authorities and the courts, is deterrence, but the offences are unlikely to be achieving this goal. The penalties imposed are of no more than moderate severity and this is then heavily discounted by the low probability that such a penalty will be imposed (which is generally considered the stronger factor). Together, this means that the expected cost of the offences is minimal, and as such it will only take a correspondingly low expected benefit from offending before a rational regulatee will breach the Act.

If deterrence is indeed the primary goal, then we need to understand what is keeping the severity of the penalties, and the certainty they will be imposed, to these levels. This is the topic of this chapter, and it is considered from the perspective of the three parties that can influence these factors—the local authorities, the public and the courts—again dividing the discussion into certainty and severity.

M. Wright, *Responding to Environmental Crimes*, Palgrave Studies in Green Criminology, https://doi.org/10.1007/978-3-030-89250-0_5

Why Low Certainty?

It is convenient to begin with certainty, having concluded with it in the previous chapter and because, as discussed below, the reasons why the certainty that a criminal penalty will be imposed remains low are also relevant to the approach local authorities take to sentencing, and so the severity of penalties. There are three questions to consider: Why do local authorities not endeavour to detect more non-compliance? Why do local authorities not prosecute more? Why does the public not step in to fill the enforcement gap, either by pressuring local authorities into prosecuting more or by bringing charges themselves? These are each addressed in turn.

Deprioritisation of Compliance, Monitoring and Enforcement

When it comes to the detection of non-compliance with the RMA, local authorities invariably have limited resources and these must be allocated to a number of different areas. This is by no means unusual for environmental regulators (see generally Gunningham and Holley 2016, 281), but is particularly acute for councils given their responsibilities go far beyond just enforcement. In particular, there are certain mandatory requirements, such as each region needing a regional policy statement and a regional coastal plan, and each district needing a district plan, which require resources to prepare (and change). After such mandatory requirements, most local authorities appreciate that compliance, monitoring and enforcement of activities undertaken pursuant to these plans is a necessary and vital part of their operations, but their focus is on growing the local economy through new businesses entering the region and expansions to existing operations. Such activities often need resource consents, so resources will next be allocated to the units responsible for processing these applications. Funding for compliance, monitoring and enforcement usually comes later (see generally Ministry for the Environment 2016, 17).

This is reflected in the relative numbers of staff involved in processing resource consents and those involved in compliance, monitoring and enforcement. For instance, in the final local authority survey (2012/2013) there were 704 fulltime equivalent (FTE) staff involved in processing resource consents and 436.7 FTE staff involved in compliance, monitoring and enforcement, and by the first year of National Monitoring System data (2014/2015) the former had increased to 753 and the latter had dropped to 374. After this, the number of FTE staff involved in processing resource consents saw an even more marked increase—to 1,030 in 2015/2016 and 1,057 in 2016/2017. On the other hand, there was first a much smaller increase in the number of FTE staff involved in compliance, monitoring and enforcement—to 409 in 2015/2016—followed by a small decrease—to 402 in 2016/2017. By this point the number of FTE staff involved in processing resource consents was, therefore, well over twice the number of FTE staff involved in compliance, monitoring and enforcement, and the latter had not even returned to the 2012/2013 levels.

It is important to keep in mind that this preference for processing resource consents over undertaking compliance, monitoring and enforcement is simply a matter of prioritisation by the local authorities, rather than a necessity. However, it is unsurprising given the Fifth National Government's (2008–2017) emphasis on the processing of resource consents. This included the Resource Management (Discount on Administrative Charges) Regulations 2010, which required local authorities to give a discount on the application fee if they do not process resource consents within the statutory timeframes, and the Resource Legislation Amendment Act 2017, which saw the introduction of a 10-working day timeframe for processing "fast track" applications. It is also tacitly acknowledged in the Ministry for the Environment's *Best Practice Guidelines for Compliance, Monitoring and Enforcement under the Resource Management Act 1991*, published under the Sixth Labour Government (2017–), which state that councils need only have sufficient access to resources to support "monitoring *high-risk* resource consents, and *most medium-risk* resource consents" (2018a, 33, emphasis added).

In a similar vein, the Guidelines state that councils need only have sufficient access to resources to support "responses to and investigation of *significant* incidents" (2018a, 33, emphasis added).

High Net Cost and Low Political Palatability of Prosecution

So far, so good, but there are still many breaches detected, so why do local authorities not prosecute these more frequently? The discussion above is a partial explanation—a prosecution may involve a number of enforcement officers and take up a large amount of time—but there is more to it than this. The starting point is that there is a lack of legislative guidance and specific direction regarding how local authorities should go about the task of enforcement. In terms of legislative guidance, the only statutory instructions in the RMA regarding enforcement are the requirements that local authorities "enforce the observance" of their plans and any national environmental standards (RMA, sections 84(1) and 44A(8), respectively). However, there appear to be no consequences if these instructions are breached and the requirement that local authorities enforce their plans (which is the more relevant of the two) is not explicitly a key element of the RMA, it instead being buried in Part 5 ("Standards, policy statements, and plans") under "Subpart 6—Miscellaneous matters" and headed "Local authorities to *observe* their own policy statements and plans" (emphasis added).

In terms of specific direction as to how to go about the task of enforcement, the Ministry for the Environment, on behalf of the Minister for the Environment, is responsible for "monitoring of the effect and implementation of [the RMA] (including any regulations in force under it), national policy statements, national planning standards, and water conservation orders" (RMA, section 24(f)). However, the Ministry has traditionally provided only weak oversight and support to councils undertaking compliance, monitoring and enforcement under the RMA (Brown 2017, 34). For instance, while the Ministry helped develop an RMA Enforcement Manual, this was more focussed on how to practically use the various enforcement tools, not when to use each tool, and it deferred to the local authorities. Further, while the Ministry prepared a 2016 report that concluded, among other things, that there is significant variation in the way councils carry out compliance, monitoring and enforcement, it did not make any recommendations for improvement (Ministry for the Environment 2016).

Following the Ministry's 2016 report (and Brown's findings), the Ministry for the Environment proceeded to issue "Draft" and then "Final" *Best Practice Guidelines for Compliance, Monitoring and Enforcement under the Resource Management Act 1991*. Intriguingly, the Ministry's draft guidelines noted that some of the RMA Enforcement Manual was out of date and these guidelines would replace the compliance, monitoring and enforcement content (Ministry for the Environment 2018b, 8), but this comment was omitted from the final guidelines (Ministry for the Environment 2018a, 9). However, it is clearly the case that some of the content in the RMA Enforcement Manual is out of date—for instance, the enforcement section still refers to the procedure for commencing a prosecution as being set out in the Summary Proceedings Act 1957, which was replaced by the Criminal Procedure Act 2011 many years ago. Regardless, the purpose of the Guidelines is to "provide advice" on how local authorities should exercise the "considerable discretion in how they fulfil their statutory functions" to "achieve the purpose of the RMA" (Ministry for the Environment 2018a, 9), and it is therefore worth considering them in more detail. This is particularly the case given that such non-binding materials can "harden up" over time (Hall 2014, 105).

It is apparent on reading the Guidelines that these largely favour approaches to enforcement that are compliance-based (as discussed in Chapter 3). For instance, reference is made to the discretion that councils have, that a "strategic approach to compliance" is required and that interventions should be targeted "according to the individual or group's willingness to comply combined with the seriousness of the breach" (Ministry for the Environment 2018a, 17–19). Specifically in terms of the use of enforcement tools, the Ministry goes on to note that local authority enforcement budgets are typically capped, the consequence of this being that councils "may have to be selective about which prosecutions they take" (Ministry for the Environment 2018a, 40). As such, the Ministry is only prepared to go as far as saying that the "*best practice approach* is for councils to take the prosecution if it is warranted, regardless of potential costs to council or (partial) cost recovery through fines" (Ministry for the Environment 2018a, 40, emphasis added). At the same time, the Guidelines stress the importance of lower level enforcement responses. For instance, in talking about abatement notices the Guidelines state that they should be an enforcement officer's "primary tool", which is based on the assertion that "[m]ost abatement notices issued are eventually complied with" and "so it can be a cost-effective way of achieving positive behaviour change" (Ministry for the Environment 2018a, 88).

Turning then to the individual enforcement policies of councils, while not every local authority has one—Brown found that, as at 2016, only 37% of local authorities (26% of district councils, 36% of city councils, 73% of regional councils and 67% of unitary councils) had such a policy in place (2017, 46)—those that do also typically subscribe to compliance approaches to enforcement. For instance, as with the Ministry's Guidelines, reference is frequently made to the wide range of tools available, the broad discretion local authorities have (Environment Canterbury 2010, 3), and the need for "proportionality" between the breach and the enforcement response (West Coast Regional Council 2013, 2; Waikato Regional Council 2016, 5). This is unsurprising, given local authorities had considerable input into the drafting of the Guidelines, with Council staff on the "panel of CME experts" and staff from eight local authorities reviewing drafts of the document (Ministry for the Environment 2018a, 8).

When it comes to the circumstances in which prosecution will be used, there are a variety of approaches set out in the individual enforcement policies. Some councils are quite brief—for instance, the Bay of Plenty Regional Council in its *2016/2017 Regulatory Compliance Report* notes that "[p]rosecutions are generally reserved for more serious offences where significant environmental effects have occurred, or where there has been repeated serious non-compliance" (2017, 53). Similarly, the Taranaki Regional Council's *Enforcement Provisions and Procedure under Resource Management Act* (sic) states that the Council's process of enforcement is a "staged one of assistance and warnings in the form of inspections, Abatement Notices and Infringement Notices, leading only if necessary, to Prosecution", although prosecution may be "considered in the first instance if the nature, circumstance and effects of an offence are serious enough to warrant it" (2017, 47). On the other hand, some local authorities list a large number of relevant matters—for instance, the actual and potential adverse effects on the environment, whether the breach was as a result of deliberate, negligent or careless action, whether there was a failure to act on prior instructions, advice or notice, and so on—and then simply say that a "prosecution may be considered appropriate when the factors listed…indicate that the matter is sufficiently serious to warrant the intervention of the criminal law" (Marlborough District Council 2018, 16).

Regardless of the exact approach stated, what is common is that these all point to prosecution being used infrequently. Indeed, if councils are using formal warnings for a "minor to moderate incident" (Environment Southland 2017a, 19), or infringement notices for "*serious* non-compliance offences which do not warrant further action" (Bay of Plenty Regional Council 2017, 52, emphasis added), then for some councils the threshold for prosecution is very high indeed. Such an approach is sometimes referred to in the literature as prosecuting as "last resort" (Hawkins 2002), but it is probably more accurate to say here that local authorities will prosecute only as a "last resort" (that is, when all other options have failed) or where the damage caused is suitably severe, and in many instances both are in fact present.

The distinction between the two circumstances in which local authorities will prosecute, and the ramifications, will be returned to in the next chapter, but for now it is necessary to explain the contradiction that has emerged. As noted in Chapter 4, local authorities say they are prosecuting to deter, which relies on them frequently bringing criminal charges (and obtaining high penalties), yet they have adopted compliance-based approaches to enforcement that favour prosecuting rarely (and the Ministry for the Environment actively encourages this). It is suggested that the reality of the situation is that local authorities have adopted compliance approaches because they have a lower "net cost" (cost less recovery) than other methods, and they are more politically palatable.

Starting with the cost involved, as noted above most councils have caps on their enforcement budgets, which gives a major incentive to deal with matters informally (either by way of a verbal warning or letter) or at a lower level (say, an abatement notice or infringement notice), given the lesser expense associated with these responses. This lower cost is obvious in the case of informal measures, but infringement notices and abatement notices are also quick and inexpensive to prepare. This is because there are standard forms for each, they are issued by an enforcement officer who will not need to consult with an enforcement decision group (and in some circumstances, not even with his or her manager) (Brown 2017, 47) and legal advice will also typically not be required. If the local authority is ever challenged it can simply cancel the abatement notice, or decline to pursue the infringement notice to hearing.

On the other hand, prosecuting is time-consuming and expensive. The Ministry for the Environment's "Draft" *Best Practice Guidelines for Compliance, Monitoring and Enforcement under the Resource Management Act 1991* stated that the "significant costs of taking prosecutions can therefore deter councils from taking this action" (Ministry for the Environment 2018b, 39), but this sentence was omitted from the final guidelines, the Ministry preferring to highlight that "[t]he costs and risks of taking prosecutions can, however, be substantially reduced if the prosecution is run well, and councils engage specialist investigative staff" (Ministry for the Environment 2018a, 40). However, the comment from the draft Guidelines is nevertheless true. As set out in the 2016 report by the Ministry for the Environment, the costs of prosecution can be "considerable", in that:

> ...a simple legal opinion prior to filing charging documents may cost around $3000-$5000. A case involving an early guilty plea by a single defendant facing a single charge may be resolved with legal expenses to council of less than $10,000. If charges are defended and the matter is particularly complex, however, then legal expenses can easily run to six figures. (2016, 25)

Indeed, even undefended cases can cost a local authority hundreds of thousands of dollars. For instance, in *Bay of Plenty Regional Council v*

Mobil Oil New Zealand Ltd [2016] NZDC 8903 the sentencing Judge, after imposing a fine of $288,000 for a guilty plea at the first available opportunity to a single charge of discharging a contaminant into water, noted that the Council's actual costs in relation to the prosecution had been over $300,000. Further, it took just over a year from the incident to sentencing, meaning that Council staff will have been tied up in the proceeding, in one form or another, for this time.

Related to the cost is the tendency that councils have to be risk averse when it comes to prosecuting. Although the *Solicitor-General's Prosecution Guidelines*, which local authorities are instructed to apply (Ministry for the Environment 2018a, 70) and which they typically say they do apply (see, for instance, Environment Southland 2017a, 26), only state that the evidence required in order to commence a prosecution be "sufficient to provide a reasonable prospect of conviction" (Crown Law Office 2013, 5.1.1), councils appear to apply a heightened standard, somewhat akin to "no chance of failure". This means that, when they decide to prosecute, they will likely spend more time and money ensuring that the case is "watertight", leaving less room in the enforcement budget for other cases. There are various reasons for this risk aversity, including that because councils prosecute relatively rarely being successful means a lot to them from a reputation perspective (that is, they like to promote what has been called a "posture of invincibility" (Hawkins 1984, 190)), but the most important for our purposes is that local authorities are concerned about having costs awarded against them if a prosecution is unsuccessful. For instance, in *DNS Forest Products (2009) Ltd v Gisborne District Council* [2020] NZHC 2437 the Judge recorded the advice from the prosecution lawyer that leave had been sought to withdraw the charges against two co-defendants "because of the Council's potential exposure to a large costs award in the event the prosecution was unsuccessful" (at [16], footnote 4).

The next point is that it is not only cost, but also recovery that is important to local authorities when it comes to enforcement. For instance, the West Coast Regional Council's *Enforcement Policy* states first that the Council "will [endeavour] to undertake its duties in a clear and speedy manner that gives effect to the purpose and principles of the RMA while minimizing costs to the ratepayer" (2013, 3). Then, after noting that it "must often make decisions associated with the allocation of finite resources to areas of highest priority", it states that "[i]n order to ensure this principle is not compromised through a lack of funds, the Council intends to recover the costs associated with enforcement where possible".

However, a local authority's ability to recover its costs relating to compliance and monitoring is variable. On the one hand, local authorities are permitted to "fix charges" on resource consent holders for "the carrying out by the local authority of its functions in relation to the administration, monitoring, and supervision of resource consents" (RMA, section 36(1)(c)), and a local authority that issues an infringement notice gets the full fee. On the other hand, the recovery of costs associated with permitted activity monitoring appears to be limited to situations where the council has, in a bylaw, prescribed fees or charges for inspections, and then undertaken such an inspection under s 332(1) of the RMA (Local Government Act 2002, section 150(1)(b)). Further, there are no specific recovery provisions relating to complaints, although it would appear that some councils simply invoice alleged transgressors expecting, or perhaps hoping, they will pay without legal challenge (Brown 2017, 43–44; West Coast Regional Council 2013, 3; Environment Southland 2017b, 55).

It is, of course, the case that, pursuant to section 342 of the RMA, local authorities get (at least, but not usually more than) 90% of any fines imposed in any prosecutions they bring. However, there is no guarantee that this will cover the cost of prosecuting, or that it will even be recovered (Ministry for the Environment 2018a, 40). In terms of the former, even the fine of $288,000 in *Bay of Plenty Regional Council v Mobil Oil New Zealand Ltd* [2016] NZDC 8903, let alone 90% of it, reportedly did not cover the costs of prosecuting (at [59]). And in terms of recovery, the Ministry for the Environment has noted that there "are a number of substantial RMA fines that have never been recovered by the Ministry of Justice and the fines have been 'remitted', often due to a company going into liquidation or receivership, or the offender absconding" (Ministry for the Environment 2016, 25).

That the net cost of enforcement mechanisms appears to be a significant factor in the approach of local authorities is no real surprise.

As the Ministry for the Environment concluded in 2016, "Councils' decisions to prosecute are often influenced by the costs of taking prosecutions, which are significant, and often not recovered completely through fines" (Ministry for the Environment 2016, 41). Further, a "[l]ack of resources for compliance function" and the "[h]igh cost of enforcement processes" were, respectively, the two biggest challenges identified by respondents to Brown's survey of RMA compliance, monitoring and enforcement by local authorities (2017, 54). However, it leads one to recall what Ashworth has noted when discussing compliance versus deterrence approaches to enforcement, namely that:

> If it transpires that the real argument is that the various inspectorates have insufficient resources to mount more prosecutions, and therefore use the compliance approach partly for cost reasons, then we must accept that the debate has shifted to a rather different terrain. (2000, 248)

The cost and recovery of the various enforcement mechanisms is related to the final factor that has led to the adoption of compliance approaches, and that is (what could be termed) "political palatability". What is meant by this is that informal and low-level enforcement measures are often seen by the enforcement units of local authorities as more acceptable to elected officials, with councillors only considering prosecution to be appropriate when there is no other option. The starting point for this argument is that monitoring and enforcement is sometimes viewed in a dim light by regulated parties. This is, of course, not universal—many people and businesses, typically those that are complying, understand the importance of it to, at least, maintain a level playing field, even if they are sometimes sceptical as to the positive effects on the environment that it has—but it is often the case that many in the community are resistant. For instance, in terms of monitoring, reports that the Waikato Regional Council was reinstating the aerial surveillance of farms drew a backlash from some farmers (Smallman 2018). And in terms of enforcement, the sentencing Judge in *Waikato Regional Council v Hold The Gold Ltd* DC Hamilton CRN14019501673-74, 19 December 2014 expressed concern at the criticisms the community had made of the Council in deciding to prosecute the defendants, and tried to ameliorate these by setting out in detail the Council's responsibilities in enforcing the RMA. Indeed, in this case despite the prosecutor and defendant both submitting that the latter should simply be convicted and discharged, the sentencing Judge disagreed and convicted the defendant and fined it $16,875, noting that "this decision will in all likelihood not be welcomed by all of those in the dairy farming industry" (at [56]).

This is important because certain sectors of the regulated community are often powerful and have influence with councils, and may even be represented (and sometimes in the majority) at the councillor level. It is no surprise then that there is a view that elected officials can, at times, see formal enforcement action as "undesirable" (Brown 2017, 37), with some councillors even running for office on the basis that, if elected, they will amend the council's enforcement practices (Ministry for the Environment 2016, 27). Further, in some local authorities councillors have been involved with specific enforcement decisions (Ministry for the Environment 2016, 26–27; Brown 2017, 48), despite the inappropriateness of this (Office of the Auditor-General 2011, 59–68). And even when they are not formally involved there is evidence that councillors can make their dislike for prosecution known in other ways (Office of the Auditor-General 2019, 60). For instance, the minutes of the Resource Management Committee of the West Coast Regional Council from one meeting in 2017 record a councillor giving his view that a matter that resulted in a successful prosecution should instead have been "sorted out on site", and expressing "his concern at what this could cost" (West Coast Regional Council 2017, 3). While another councillor supported the prosecution, noting that "Council's last option is legal proceedings and in this case Council has gone through exactly the right processes", the fact that councillors are openly discussing specific cases—even concluded ones—means that compliance staff will therefore be aware of their views on enforcement generally, and this may affect their decision-making in future cases.

These views on enforcement action are likely to then filter down to the compliance, monitoring and enforcement unit of the local authority. For instance, there is evidence that, at least in some small councils, staff feel pressure to adopt informal measures to deal with non-compliance (Ministry for the Environment 2016, 28–29). However, in some cases warnings or letters will not be enough due to the extent or nature of the breach, and it is here that infringement notices and abatement notices become such powerful tools for local authorities. This is because they "count" as a formal enforcement action, in just the same way that prosecutions do, and so can give the impression that the council is taking a matter more seriously than it perhaps is.

Impracticality of Public Participation

Before moving on to consider the reasons why the severity of the penalties is no more than moderate, it must be asked why, if local authorities are treating detected breaches too leniently, the public is not, either individually or in groups, seeking to change this. There are two main ways it could do so, by pressuring councils to prosecute instead of using informal or low-level means or by undertaking private prosecutions. However, it is suggested that neither of these options is practical, for three (related) reasons.

First, many RMA breaches have no direct human victim, and as such there is often no one that the local authority must explain itself to if it decides not to prosecute, or anyone with a sufficient personal interest to bring proceedings themselves. In a similar vein, any complaints that are made are likely to be from a member of the general public, so they too can be expected to be less inclined to follow up with the council regarding what—if any—action was taken, or to take action themselves. Importantly, there are no arrangements, such as are found in the Health and Safety at Work Act 2015, whereby people can express an interest in a case (section 142) and the regulator has "first right of refusal" to bring a prosecution (section 143), but if the regulator chooses not to do so they must notify those with an expressed interest and they may take a private prosecution (section 144). Nor are there any non-government organisations that have enforcement as a focus. While a variety of organisations are acutely interested in RMA non-compliance, their focus has tended to be on (what they consider to be) broader trends and how councils should change their actions (see, for instance, Forest & Bird 2018), rather than intervening in specific cases.

Second, although there is sometimes an immediate public notification following a (suspected) RMA breach, such as a news article or press release from a council, the actual investigation itself, and its outcome, is typically not reported unless the end result is a successful prosecution. Put another way, local authorities are unlikely to proactively report on serious incidents that get resolved by low level measures or informally, such as where the party at fault has remediated the damage and no action is taken. This lack of notification of the outcome of any investigation is important because, by the time it becomes known to the public, it may be too late for the local authority to change its mind, or for an individual to prosecute the transgressor themselves. The reason for this is that, until 2020, the limitation period for bringing a prosecution ended "6 months after the date on which the contravention giving rise to the charge first became known, or should have become known, to the local authority or consent authority" (RMA, section 338(4)). This has now been increased to 12 months, but it still means that a situation could arise where a local authority knew (or should have known) about a contravention, and because too long has passed a member of the public, despite only recently learning about the breach, cannot prosecute.

Third, the three things that are generally considered critical to the enforcement of environmental laws by the public—evidence-gathering abilities, low cost and the ability to recover these costs (Naysnerski and Tietenberg 1992)—are lacking. Private individuals do not have access to the RMA's inspection provisions, with the power of entry for this purpose being limited to enforcement officers (RMA, section 322). Further, the cost to an individual to bring a prosecution is likely to be prohibitively expensive (Ministry for the Environment 2018a, 71). Finally, and perhaps most galling of all, even if an individual did bring a prosecution and was successful, his or her only avenue for recovery of costs incurred would be the Costs in Criminal Cases Act 1967 (which, as discussed in the next section, is very limited), as section 342 of the RMA (which provides that the fine is returned to the party bringing the proceedings) is restricted to local authorities. Put shortly, local authorities have an effective monopoly on prosecution, and the wording of section 342 suggests that central government is quite happy for that to be the case.

Why No More Than Moderate Severity?

We can now consider the reasons why the penalties imposed on those that have been successfully prosecuted for breaching the RMA are of no more than moderate severity. There are, again, three questions to consider: Why are fines still only modest? Why are other, more restrictive, penalties not being imposed? Why do local authorities not "use" convictions or leverage prosecutions to create informal penalties against offenders? These are each addressed in turn.

Restrictions on Fine Levels

Let us start with fines, which, as noted in the previous chapter, are the predominant sentence imposed by the courts for RMA offending. It is, of course, undoubtedly true that average fine levels have increased, and this is especially the case since the 2009 amendments to the maximum penalty. However, at $36,268 in 2016/2017 average fines are still only, at most, approximately 12% of the maximum penalty the Court can impose, and were accordingly described in Chapter 4 as modest.

Further, in light of the analysis in the previous section, the modesty of these fine levels can now be put another way. The Sentencing Act 2002 provides that the maximum penalty (here imprisonment for a term not exceeding 2 years or a fine of $300,000 for natural persons and a fine of $600,000 for non-natural persons) should be imposed for offending "within the most serious of cases for which that penalty is prescribed, unless circumstances relating to the offender make that inappropriate" (section 8(c)). Similarly, the courts must impose "a penalty near to the maximum prescribed for the offence if the offending is near to the most serious of cases for which that penalty is prescribed, unless circumstances relating to the offender make that inappropriate" (section 8(d)). Average fines per prosecution of around $35,000 suggest that offending is typically (much) closer to the least serious end of offending, yet this is at odds with the approach taken by local authorities, of prosecuting only as a "last resort" (that is, when all other options have failed) and/or where the damage caused is severe.

If Environment Judges have accepted that deterrence is (usually) the primary purpose of RMA sentencing—and it was argued in the previous chapter that there is considerable evidence that they have—the question that needs to be asked is why they continue to impose fines that are only modest, especially when so few prosecutions are brought. In the literature, low fines are sometimes attributed to the absence of specialist Judges (Hampton 2005, 39–40; Cabinet Office (UK) 2006, 14), but this is obviously not the case here, with all prosecutions heard by an Environment Judge (unless otherwise directed by the Chief District Court Judge). Rather, it is argued that fine levels are only modest because Environment Judges are required to follow orthodox criminal sentencing practice—determining the sentence considered appropriate for the particular offending (the combination of features) for an adult offender after a defended trial (the "starting point") and then making adjustments for personal aggravating or mitigating factors—which limits these penalties in two ways. First, it means that the judges cannot apply deterrent models when calculating penalties, instead being forced to apply a desert-based approach that, in the context of the RMA, tends to restrict starting point fine levels. Second, the judges believe that they must give defendants a range of discounts from, and are rarely able to uplift, the starting point in determining the end fine.

Beginning with the first of these arguments, if deterrence is the primary purpose of sentencing RMA offenders then this suggests that sentences should be driven by deterrent models (Yeung 1999), the two most common of which are optimal deterrence and absolute deterrence (Yeung 2004, 65–66; Bilmes and Woodbury 1991). Optimal deterrence weighs up the social costs and social benefits of non-compliance, only seeking to deter conduct for which the former is greater than the latter. Penalties are based on the harm caused (including enforcement costs), with the fact that the offender may have received a large benefit from offending, and his or her ability to pay, being irrelevant. On the other hand, absolute deterrence takes the position that all unlawful conduct should be prevented. There is no need to consider social benefits, because it is assumed there are none (or they are always outweighed by the social costs) (Yeung 1999, 449). The focus when setting the penalty is on the benefit to the transgressor, rather than the harm caused or the enforcement costs incurred, although the benefit obtained is only the minimum penalty, as over-deterrence is not a problem in the way that it is in the optimal deterrence model (Yeung 2004, 65–66). In both models, the probability of detection and successful prosecution needs to be taken into account. If this is less than 100% (which, in reality, it always will be), the penalty imposed needs to be increased by a commensurate amount (Yeung 1999, 449–450)—by way of simple example, if the penalty according to the harm caused or benefit obtained is $100, but the probability of detection and successful prosecution is only 50%, the penalty should be $200 ($100/0.5).

However, there is little evidence that the courts apply either model when sentencing RMA offenders. It is, of course, the case that harm is one of the main factors considered by the courts in setting starting points, but there is typically no value specifically attributed to it, the courts instead often falling back on the comment that harm is difficult to quantify. Similarly, although the higher courts have made reference to taking into account the benefit to the offender, with, for instance, Miller J in *Thurston v Manawatu-Wanganui Regional Council* HC Palmerston North CRI-2009-454-24, 25 & 27, 27 August 2010 stating that the "Court may prefer to focus on deterrence by eliminating the offender's gains, which are more easily quantified, before considering whether any further penalty is warranted to reflect environmental harm" (at [46]), this approach does not appear to have been adopted by Environment Judges. Rather, there is generally no effort made to calculate the defendant's gains from offending or, if there is, these gains bear no relevance to the starting point adopted. The probability of detection and successful prosecution is rarely mentioned; indeed, the court is much more likely to uplift the penalty where many prosecutions are brought (on the basis that there is a problem in the region that needs attention, the prime example being dairy effluent discharges), rather than in those difficult to detect and prosecute areas such as land use (a point which is returned to in the next chapter).

To understand the reason why deterrent models have not been adopted, despite Environment Judges frequently referring to deterrence being the main purpose of sentencing, it is necessary to consider a series of RMA appeals. The first of these is the High Court's decision in

Machinery Movers Ltd v Auckland Regional Council [1994] 1 NZLR 492 (HC) in which, as noted in Chapter 4, a "Full Court" (two judges) was convened to consider the proper approach to RMA sentencing. In this case, the Court considered, and largely adopted, the reasoning in the Canadian first instance sentencing decision of *R v Bata Industries Ltd* (1992) 9 OR (3d) 329, 7 CELR (NS) 293 (Ontario Court (Provincial Division)). There, Ormston Prov J had set out relevant factors in assessing the severity of an offence, including the nature of the environment affected, the extent of damage inflicted, the deliberateness of the offence and the attitude of the accused. The Judge in *Bata Industries* also observed that, for corporations, the Court should consider the size, wealth, nature of operations and its power, the extent of attempts to comply, remorse, profits realised and the criminal record or other evidence of good character. Ormston Prov J then added that:

> When a court imposes a penalty in respect of an environmental offence, the level should reflect the gravity of that *particular* offence and leave room for the most serious of circumstances. The harshest sentences ought to be reserved for the worst possible factual situation.

Putting to one side for the moment the specific factors relevant to corporations, which other than profits realised go to aggravating and mitigating factors personal to the offender (and which are dealt with below), what is apparent is that Ormston Prov J largely mandated (and the New Zealand High Court adopted) an orthodox, desert-based, enquiry to sentencing. In particular, and as noted in Chapter 3, desert theory suggests that the sentence should be proportionate to the seriousness of the breach, with seriousness primarily made up of harm (here, the environment affected and the extent of damage inflicted) and culpability (here, the deliberateness of the offence and the attitude of the accused) (Von Hirsch et al. 2009, 104). Also consistent with desert theory, no consideration is given to matters such as the costs of enforcement or the probability of detection and successful prosecution.

The primacy of a desert-based approach to setting penalties was then reinforced nearly 10 years later, with the enactment of the Sentencing Act 2002 and its adoption by the higher courts in RMA cases. This is because, although the Act sets out all rationales of punishment in section 7 as available options, the principles of sentencing in section 8 suggest that, when actually imposing the penalty, proportionality (to seriousness) should be the overriding concern (Roberts 2003, 259). When the High Court in *Selwyn Mews Ltd v Auckland City Council* HC Auckland CRI-2003-404-159-161, 30 April 2004 considered the interface between the Sentencing Act 2002 and *Machinery Movers Ltd*, it noted that the sentencing principles established in the latter continue to have application, but are not comprehensive and must now be read in light of the Act. This was confirmed by the Court of Appeal in *R v Conway* [2005] NZRMA 274, with the Court observing that the proper approach was to apply the two statutes "in harmony" (at [60]). Finally, RMA sentencing was then explicitly brought within the criminal orthodoxy in *Calford Holdings Ltd v Waikato Regional Council* (2009) 15 ELRNZ 212 (HC). In this case, the High Court held that Environment Judges are to sentence RMA offenders in exactly the same way as for any other criminal offence, which means, as referred to above, setting a starting point for the offending—"the sentence considered appropriate for the particular offending (the combination of features) for an adult offender after a defended trial" (*R v Mako* [2000] 2 NZLR 170 (CA) at [34])—and then making adjustments for personal aggravating or mitigating factors, including a discount for a guilty plea (if any).

However, an orthodox approach to sentencing that uses desert-based principles is likely to be restricting the starting points that are being adopted. The issue is not so much with the "harm" limb, as while (as referred to above) there are issues in quantifying this the courts are generally prepared to make "an allowance for harm…on the assumption that any given offence contributes to the cumulative effect of pollution generally" (*Thurston* at [41]), and the judges are well aware that it is "routine" for defendants to:

> … seek to minimise the impact of the offending on the environment by either referring to a lack of proof of actual effects on the environment or arguing that within the context of the whole environment purported to be affected, the impact is minimal. Such submissions tend to ignore or seek to minimise the definition of effect in s 3 of the RMA and its reference

> to potential and cumulative effects and the purpose and approach of the RMA to effects on the environment. (*Bay of Plenty Regional Council v Withington* [2018] NZDC 1800 at [34])

Rather, the problem is more when the courts come to assess culpability, the other main limb of the desert enquiry. As noted in Chapter 2, *mens rea* is not a necessary part of the primary offences—section 341(1) of the RMA provides that it is irrelevant to contraventions (or permitting the contravention) of sections 9, 11, 12, 13, 14 and 15 whether or not the defendant intended to commit the offence—and the courts have held that it is also not required in the two other most common offences, breach of an enforcement order and breach of an abatement notice. Therefore, it is a matter of argument come sentencing (Sentencing Act 2002, section 8(a)), but because of the high percentage of guilty pleas in RMA cases—typically between 80 and 90% (Ministry for the Environment 2013, 7)—this argument is typically based on a summary of facts, rather than evidence seen and heard by the Judge. This document then becomes a "battleground", with defendants often accepting that an offence has been committed, but taking issue over the facts alleged by the prosecutor due to the implications this has for, among other things, culpability.

The defendant is at something of an advantage here for two (related) reasons. The first is that one of the motivations for removing culpability from an offence is that the defendant is typically in possession of the relevant facts relating to whether or not (and how much) it is at fault, and this may not be easily ascertainable by the prosecution. As such, the prosecution may well be on the back foot in any negotiations, or have to spend a considerable amount of time and money (for instance, on experts) obtaining evidence of this. Second, the prosecutor will be anxious to avoid a disputed facts hearing (which occurs when a relevant fact or facts cannot be agreed (Sentencing Act 2002, section 24)), given that this will further add to the time and cost involved. While a defendant will also want to avoid a disputed facts hearing, for the same reasons, the incentive may be less, given it is the prosecutor that bears the onus of proving beyond a reasonable doubt the existence of any disputed aggravating fact (Sentencing Act 2002, section 24(2)(c)). The end result can, therefore, be that the prosecutor tempers the summary of facts in order to get agreement, and sentencing will accordingly proceed in a more favourable light for the defendant (see generally de Prez 2000).

Whether for this reason or not, whatever culpability the prosecution can attribute to the defendant is likely to only be carelessness, or recklessness at worst. For instance, in the fourth and final prosecution study published by the Ministry for the Environment, 58% of cases were in the "careless" to "high level of carelessness/negligence" spectrum, with 10% having an "element of deliberateness" and 15% deliberate, and 8% accidental (and 6% no finding) (Ministry for the Environment 2013, 18). Similarly, in the 2016/2017 sentencing cases reviewed most cases were assessed by the Judge as being in this careless to reckless spectrum, with a small number either side (that is, being largely accidental but caught by liability without fault or, at the other end of the spectrum, intentional). This means that the effective maximum penalty is typically reduced, as space is left at the top for the "most serious of cases for which that penalty is prescribed" (Sentencing Act 2002, section 8(c) and (d)). A paradigm example of this is *Maritime New Zealand v Daina Shipping Company* DC Tauranga CRI-2012-070-1872, 26 October 2012, where despite this case being "New Zealand's worst maritime environmental disaster" (at [2]) with "hundreds of tonnes of oil" and "hundreds of containers and their contents, including dangerous goods" (at Endnote ii) being discharged into the ocean, and the total cost to the Crown (as at October 2012) being approximately $47 million, a starting point of $450,000 (three-quarters of the maximum) was adopted because the Judge considered he had (at [9]) "to leave room for those cases where there has been deliberateness, recklessness, and/or a high level of carelessness by the person being sentenced".

More generally, the issue with a desert-based approach to RMA sentencing is that, once the assessment is made as to the gravity of the offending, the starting point adopted needs to (broadly) fit in with that adopted in similar cases (section 8(e) of the Sentencing Act 2002), and historically even serious environmental breaches have attracted relatively low sentences (Resource Management Law Reform Core Group 1988, 47). As such, while starting points can increase—and have done so—this only happens incrementally, as individual cases that attract inconsistently high starting points are liable to be overturned on appeal if the end result is "manifestly excessive". While it appeared that the introduction of the RMA and the $25,000 fine upheld in *Machinery Movers Ltd* might signal an immediate and significant increase in penalties across-the-board, this result is best seen as an outlier, given the average total fine per case during the first ten years of the RMA (when this decision was issued) was only $6,500 (Ministry for the Environment 2002, 8). Ultimately, the desert-based approach has meant that it has taken a long time for starting points even to reach where they are now.

The problem of restricted starting points is then compounded by Environment Judges considering themselves required, based on criminal law orthodoxy and the provisions of the Sentencing Act 2002, to give a number of discounts for mitigating factors personal to the offender, but rarely considering themselves able to provide an uplift for aggravating factors. Beginning with the latter, the predominant factor that can lead to an increase in the starting point is relevant previous convictions (Sentencing Act 2002, section 9(1)(j)), but given prosecution is used sparingly most offenders will only have council warnings, infringement notices or abatement notices as part of their history. Indeed, with the advent of the formal diversion schemes that are offered by some local authorities (see Chapter 6), this will only increase. This is because it has been held that a defendant that has previously gone through such a process and offended again will not receive an uplift to the starting point, as they will not have received a conviction upon completing the diversionary scheme (*West Coast Regional Council v Johnson Brothers Transport (2006) Ltd* [2018] NZDC 22057).

On the other hand, the courts generally recognise a number of discounts from the starting point adopted. These include for previous good character (Sentencing Act 2002, section 9(2)(g)), remorse shown by the offender (Sentencing Act 2002, section 9(2)(f)), remediation/restoration efforts and upgrades to infrastructure to prevent the offending happening again (although not on a "dollar-for-dollar" basis). Most importantly, the defendant will receive a discount of up to 25% for entering a guilty plea (see Sentencing Act 2002, section 9(2)(b) and *Hessell v R* [2010] NZSC 135, [2011] 1 NZLR 607). The effect of these discounts (and lack of uplifts) can best be seen in the 39 cases reviewed in the 2016/2017 financial year, whereby aggregated starting points of $2,079,500 were reduced by just over a third to aggregated fines of $1,378,198.

Finally, section 40(1) of the Sentencing Act 2002 provides that, in determining the amount of a fine, the Court must take into account the financial capacity of the offender. In a limited number of cases, the courts have relied on this provision to increase the fines, but it would seem that they are only prepared to do so when the defendant is a very large corporation, and as such it has limited effect. For example, in *Bay of Plenty Regional Council v Fonterra Limited* DC Tauranga CRI-2015-087-385, 27 July 2015 there was a 50% uplift to the starting point, mainly for the size and wealth of the defendant but also for previous offending, and in *Mobil Oil New Zealand Ltd* there was an uplift of one-third to the starting point, due to the size and wealth of the defendant. However, such uplifts are unlikely to make much difference for these companies—the parent company of Fonterra Limited (Fonterra Co-operative Limited) had a net profit after tax of $506m in the year that ended four days after Fonterra Limited was sentenced, while the parent company of Mobil Oil New Zealand Ltd earned US$16.2 billion in the previous year. Instead, section 40(1) more typically has the effect of decreasing the amount payable, with such reductions sometimes being substantial as the Court is generally unwilling to impose a penalty that will put the offender out of business (see *Machinery Movers* at 509). Indeed, if the company is already in liquidation then the penalty may get reduced to nothing (see, by way of example, *Tasman District Council v Hunter Laminates 2014 Ltd (in liq)* [2018] NZDC 17605).

It is here, however, that the courts appear to have taken a step to limit this effect of section 40(1), by accepting that defendants can insure against RMA fines. For many years this had been allowed by the courts without question (but see Jurgeleit 1996), and on occasion defendants had even been given discounts to reflect that they "responsibly obtained insurance in respect of the payment of the fine" (*Canterbury Regional Council v Canterbury Greenwaste Processors Ltd* DC Christchurch CRI-2012-009-9818-9820, 24 April 2013 at [105]). However, the ability to insure against fines was argued in the 2018 case of *Bay of Plenty Regional Council v Whitikau Holdings Ltd* [2018] NZDC 3850 on the basis that it would seem to contradict the maxim *ex turpi oritur causa non oritur actio*—"an action does not arise from a base cause" (*Burrows v Rhodes* [1899] 1 QB 816). The issue for the Court was an unenviable one—either prohibit insurance against fines and be left to reduce heavily the end point for impecunious (corporate) defendants, or allow insurance against fines, which means the defendant will only need to pay the insurance excess. Ultimately, the Court opted for the latter, on the basis that the sentencing Judge did not believe, in the absence of an express statutory provision, she had the ability to declare the insurance policy void. This meant that even where the defendant is impecunious the deterrent effect of the sentence will be preserved, although it will be suggested, at the end of the next section, that this decision can be explained in another way.

A Preference for "Compensatory" Sentences

If the courts are, as just argued, constrained from imposing sterner fines (other than on a slowly incremental basis), the next point to consider is why they are not imposing other, more restrictive, *types* of sentence, in order to increase deterrence. As noted in the previous chapter, the alternative sentences that a court can impose for RMA offending are non-monetary sentences for natural persons (community work, supervision, intensive supervision, community detention, home detention and imprisonment), "compliance-focussed" enforcement orders, "punitive" enforcement orders and resource consent reviews, but the vast majority of cases are resolved by way of a fine (with, in some instances, a "compliance-focussed" enforcement order). While for natural persons there is a presumption in favour of a fine in the Sentencing Act 2002, there is much scope for courts to move beyond this, such as if it would not achieve a purpose (or purposes) of sentencing (which includes deterrence), the application of any of the sentencing principles would make it inappropriate, or it would be "clearly inadequate in the circumstances" (Sentencing Act 2002, section 13(a), (b) and (d)). In terms of non-natural persons, it is, of course, the case that the traditional non-monetary sentences are not available, but this does not explain the lack of consideration for a "punitive" enforcement order, such as was made in *Bay of Plenty Regional Council v Hood* [2017] NZDC 25156 and encouraged in *Machinery Movers Ltd*, or a resource consent review, which can result in the cancellation of the consent. Put shortly, sentencing an individual to a more restrictive penalty than a fine, or ordering a company to temporarily or permanently cease operating, would seem to be a far greater deterrent than a monetary penalty, yet this is rarely asked for or ordered. It is suggested, in this section, that local authorities prefer fines (sometimes with a compliance-focussed enforcement order) because these penalties are "compensatory", whereas the other sentences are not, and the courts typically follow the wishes of the local authority in terms of the type of penalty imposed (but not necessarily the quantum).

On the face of it, referring to a fine as a "compensatory" sentence seems misplaced. However, from the perspective of the local authority this is exactly its effect, given section 342 of the RMA. This provision, which operates automatically, requires that at least 90% (but usually no more) of the fine imposed be paid to the local authority that commenced the proceeding. In essence, therefore, the local authority is being compensated for some or all of (or even more than) the cost it has incurred in investigating and prosecuting the matter.

Local authorities consider this compensation to be important for two reasons. First, as noted when discussing certainty, prosecution is expensive, with even the simplest case likely to be in the tens of thousands of dollars. Complex cases, or ones where the defendant pursues an aggressive defence, will cost much more. Second, local authorities will, typically, otherwise only get a minimal amount, if any, of their costs paid. This is because costs awards are governed by the Costs in Criminal Cases Act 1967 and the Costs in Criminal Cases Regulations 1987, and these provide a scale rate of only $113 per half day for an undefended matter (typically a guilty plea) and $226 per half day for a defended case. The Court may make an order for the payment of costs in excess of that scale if "having regard to the special difficulty, complexity, or importance of the case, the payment of greater costs is desirable" (section 13(3)), but this test is not met simply because the scale is "perceived to be inadequate" (*Roberts v Northland Regional Council* [2014] NZHC 284 at [109]). In the usual course, therefore, it is only scale costs that are awarded, and even this does not always happen—in the sentencing decisions analysed from the 2016/2017 financial year, costs were awarded in 27 of the 36 cases (and never at higher than scale).

There are also, it is suggested, two reasons why local authorities are happy for "compliance-focussed" enforcement orders to be ordered in addition to a fine. One is that they are a compensatory sentence, in that the defendant is being made to address the effects of the offending or its cause (or both). The second is that they will not typically impact on the size of the fine—for instance, in *R v Fan* [2016] NZDC 7194 the sentencing Judge noted:

> [34] In this case I do not consider that an enforcement order is part of the penalty in the sense that it would simply be returning the situation to that which was lawfully permitted. For that reason I do not see the imposition of the enforcement order as a basis for reducing the fine.

On the other hand, there is no compensatory aspect when an individual receives a non-monetary sentence, and therefore such penalties are not favoured by local authorities unless the defendant can prove he or she is impecunious. For instance, in *Bay of Plenty Regional Council v Singh* [2018] NZDC 13866 the defendant pleaded guilty to five RMA charges and, when the matter was first called for sentencing, submitted that he could not pay a fine and so community work should be imposed. Despite community work being a more restrictive sentence than a fine, this was "strongly opposed" by the prosecutor (at [75]), on the basis that the local authority believed the defendant had access to funds from a trust that could be used to pay a fine, and the matter was adjourned for investigations to take place. Indeed, this preference for a fine, if the defendant can pay one, appears to hold even where the offending is serious and he or she has multiple previous convictions for RMA offending (see, by way of example, *Waikato Regional Council v Pollock Farms (2011) Ltd* [2019] NZDC 2204).

In terms of "punitive" enforcement orders and resource consent reviews, these are generally not sought by councils for two reasons. First, in the usual course, local authorities do not want people who are in breach of permitted activity rules or who breach their resource consent to stop operating; rather, they just want them to do so in accordance with the rules or the conditions of the consent (as the case may be). Second, as a penalty, these orders will often replace, or at least reduce, the fine. Indeed, this is precisely what happened in *Hood*, with the order requiring the defendant to cease farming pigs, remove all pigs and not keep any further pigs on the property, and to undertake remedial works, being imposed *instead of* a fine of $26,250.

It is important to be clear that the argument is not that local authorities are disinterested in deterrence, but rather that they want this to be achieved as a by-product of a compensatory sentence. Put another way, deterrence is not in fact the primary goal. This can be seen in the related proceedings of *Northland Regional Council v Beejay Stud Ltd* DC Whangarei CRI-2016-088-759-760, 26 October 2016 and *Northland Regional Council v Clear Ridge Station Ltd* DC Whangarei CRI-2016-088-403, 26 October 2016, and their aftermath. In these cases, fines of $135,000 and $90,000 (respectively) were imposed on the two related companies, but these were still outstanding over a year later and the latest report was that they were unlikely to ever be paid, as the farms owned by the companies had been sold and it transpired that neither had any remaining assets (Sharpe 2017). When questioned about this, a representative from the Council said it "had proceeded with the prosecution on the basis that the fines would be paid in full" and that:

> The council takes prosecutions as a last resort and expected them to set a general deterrent to other farmers and a specific deterrent to the two companies. It is disappointing that in this case, not only is there no specific deterrent, the ratepayers of Northland are out of pocket for over $50,000 in costs for the prosecutions. (Sharpe 2018)

However, it is clear that there was (in addition to the general deterrent value of the fine) a specific deterrent to the companies, in that they are no longer operating. What the representative means, of course, is that there was no specific deterrent to the owners of the companies, but the fault for this must lie with the Council as the sentencing decisions record that it withdrew the charges against the (same) director, presumably as part of a plea arrangement with the companies (as to which, see Chapter 6).

But it seems, from the final comment, that the issue was as much that the Council would not receive any compensation for the cost of prosecuting the companies in achieving this deterrence.

Before moving on to the approach local authorities take to "using" convictions and leveraging informal penalties, there is another matter to address. It has been argued that local authorities prefer compensatory penalties, but sentences are obviously imposed by the courts and therefore it is necessary to consider the approach they take. On this front, the first point to note is that, despite non-monetary sentences for natural persons, "punitive" enforcement orders and orders for a resource consent review being made by the courts, from a practical perspective they depend for their imposition on local authorities. In terms of sentencing individuals to non-monetary outcomes, while Judges are able to impose such sentences if they consider the provisions of the Sentencing Act 2002 are met, they are highly unlikely to do so if this has not been requested by the prosecution. Similarly, the making of a "punitive" enforcement order or an order for a resource consent review realistically depends on the local authority arguing for this, as it is going to involve ongoing action on the part of the council (for instance, to draft the enforcement order or to conduct the resource consent review), so the Court is unlikely to force this on an unwilling local authority.

But in any event, it is apparent that the courts agree with local authorities' preference for compensatory penalties. The starting point is the comment that is frequently made by judges in RMA sentencing cases that fines will "most often" be the appropriate sentence (see, for example, *Selwyn Mews Ltd v Auckland City Council* HC Auckland CRI-2003-404-159-161, 30 April 2004 at [40]), although there is generally a lack of explicit explanation as to why this is the case. The obvious reason why the courts would, all other things being equal, consider a fine to be the appropriate sentence is that this would be in accordance with the "polluter pays" principle, in that the offender is, literally, being made to pay for the environmental harm he or she has caused. But as was seen earlier in this section, it is rare that the environmental harm is quantified, let alone drives the starting point, and, even if it was, this would be diminished by the discounts a defendant typically receives. Further, offenders can be made to pay for the environmental harms they have caused by "compliance-focussed" enforcement orders (and, in some circumstances, an order for reparation).

As such, it would seem that the courts consider fines to "most often" be the appropriate sentence because of the compensation they provide to the local authority. For instance, in *R v Fan* Judge Harland stated, in declining a request by the defendant that the fine of $30,000 should be substituted for community work, that "the ratepayers should not have to pay Council costs, rather the costs should be to the person who caused the costs to be incurred in the first place" (at [30]). Similarly, in *R v Yealands* [2018] NZDC 4115 the sentencing Judge noted, after assessing reparation of $12,983.25, that while "[o]rdinarily" a conviction and discharge would have been the appropriate sentence to go with this, it would mean that the local authority would not receive a contribution towards its costs, and accordingly a fine of $5,000 was also imposed (at [102]). The position was summed up in *Bay of Plenty Regional Council v Merrie* [2018] NZDC 1621, where the Court stated (at [72]):

> …The prosecutor has a statutory duty under s 84 of the RMA to enforce the observance of its plan. In this case it issued abatement notices to the defendants to do so. Had those been complied with, then no charges would have been laid and no court costs incurred or sought to be recovered. The fact that each defendant contravened the abatement notices served on them and these charges followed has led to the prosecutor incurring legal costs which it seeks to recover through the fines it submits ought to be imposed on the defendants. That outcome must be within the purpose of s 342 of the RMA, which requires 90% of fines to be paid to the local authority instituting a prosecution. …

Reference was made in the previous section to insurance against fines, and the comment was made that a refusal to prohibit this preserved the general deterrent effect of such sentences where the defendant was impecunious. However, the fact that fines provide a compensatory outcome for local authorities is another way of explaining this refusal. As the sentencing Judge noted in *Waikato Regional Council v Smith* [2018] NZDC 6760, "the fine, if paid by the insurer, will ensure that the Council receives some contribution towards the costs it has incurred in bringing this prosecution" (at [73]). Without insurance, the fines to be imposed on those who have limited means would need to be reduced, and therefore the council will not get compensated (or be compensated less) for the costs it has incurred in bringing the prosecution. Having said that, in these circumstances local authorities need to be careful as to how they pitch their case, as insurance policies typically exclude cover where it was a deliberate or reckless breach of, or disregard for the provisions of, the Act. This returns us to the point made in the previous section regarding proving the culpability of defendants—local authorities are, in some circumstances, incentivised to not attribute a high level of culpability to the defendant, as if the insurer declines cover on this basis and the defendant company has gone, or goes, out of business then the local authority will receive nothing.

It must be recognised that a compensatory approach to sentencing is not necessarily unusual in a global context. For instance, White has argued that the New South Wales Land and Environment Court focuses on "reparative justice" when sentencing (2017). However, that jurisdiction is purposefully geared towards this outcome, with a separation between fines and orders that prosecution costs be paid. On the other hand, compensation under the RMA is only meant to be incidental—where a fine is imposed 90% is returned to the local authority—rather than driving arrangements to ensure that a fine is ordered (and its level preserved and its total amount paid).

No Benefit from Using Convictions and Imposing Informal Penalties

Finally, the focus by local authorities on compensatory outcomes can also explain why, as noted in the previous chapter, the formal consequences of a conviction appear low and informal penalties are variable. For instance, councils may be reluctant to "use" convictions against regulatees when considering future resource consent applications, as there is little for the council to gain by stopping a regulatee from operating in the future, but there is a "cost" in that this will be one less business contributing to the local economy. This leads one to recall what Sanders and Young have noted, namely that:

> … corporations act according to capitalist laws of economic behaviour rather than laws of due process or social justice. If forced to comply with laws which make processes uneconomic (or, more realistically, less profitable) firms will simply scale down or move their business. Regulatory agencies which genuinely care about their true clients (the workers within the industry they regulate) are thus forced into choosing between two unpalatable alternatives: corporate law breaking or reduced economic activity. (2007, 369)

In a similar vein, local authorities may not leverage prosecutions as much as they could to create informal penalties, because there is no compensatory benefit to the council in doing so. On the other hand, such an approach may impact on the council's ability to work co-operatively with the regulatee in the future and any image the local authority has cultivated of being "business friendly", as some councils describe themselves (Brown 2017, 36).

These points should not be pushed too far, as some councils will assess resource consent applications more closely when dealing with a known offender, and others regularly issue press releases outlining their latest successful prosecution. There are also uncertainties in the legislative environment regarding the extent to which local authorities can take a conviction into account when assessing a resource consent application. However, what is clear is that there is no consistent approach between local authorities and no co-ordination of such by the Ministry for the Environment. The only thing that would seem to be consistent, across-the-board, is the view that the fine is what is most important and anything else is very much secondary.

Conclusion

It can be seen from the foregoing that the ineffectiveness of the offences stems from constraints on both the courts and the public that stop them from pursuing deterrence, and choices made by councils to not in fact pursue deterrence. In terms of the latter, the choices local authorities make are, in essence, to use prosecution sparingly, for cost, recovery and political palatability reasons, and, when they do prosecute, to seek compensatory outcomes. In terms of the constraints, some of these arise from the use of the criminal law, with the practicalities of prosecuting stopping the public from taking action and sentencing orthodoxy restricting the size of fine the courts can impose. However, other constraints are caused by the choices made by the local authorities, such as the reluctance of councils to notify the public of breaches of the Act and their unwillingness to apply for "punitive" enforcement orders or resource consent reviews. As such, in addition to local authorities not pursuing deterrence, they are also stopping the courts and the public from doing so.

Indeed, the approach taken by councils to prosecutions appears distinctly "non-criminal". The *Solicitor-General's Prosecution Guidelines* contain a presumption of prosecution where there has been a contravention of the criminal law, and this "provides the starting point for consideration of each individual case" (Crown Law Office 2013, 5.7). However, when it comes to RMA breaches the presumption has, essentially, been reversed—local authorities will not prosecute unless they have no other option. In a similar vein, while seeking compensatory outcomes appears to be in line with reparation being a purpose of sentencing listed in the Sentencing Act 2002 (section 7(1)(d)), such a purpose is in fact focussed on victims of offending, not forcing the offender to repair damage to their own property (or fix the issue that caused the non-compliance) or compensating the prosecutor for the costs of bringing charges. Yet it is these latter outcomes that are given the highest priority by local authorities.

The approach taken by local authorities is in one sense understandable. With limited resources and constrained budgets it is logical that councils would prosecute only when all other options have failed. And when they do prosecute and have spent tens of thousands of dollars (if not more), and have the choice between receiving some of this back (from a fine) or not receiving any of it but the offender receiving a more restrictive sentence (for example, by being sentenced to a term of community work), then the first option again makes sense. But in these circumstances it is disingenuous to claim that the primary purpose of the offences is deterrence. And further, as will be discussed in the next chapter, the approach taken by local authorities can lead them to act inappropriately when making prosecution decisions.

References

Ashworth, Andrew. 2000. Is the Criminal Law a Lost Cause? *Law Quarterly Review* 116: 225–256.

Bay of Plenty Regional Council. 2017. 2016/2017 Regulatory Compliance Report.

Bilmes, Jack, and John Woodbury. 1991. Deterrence and Justice: Setting Civil Penalties in the Federal Trade Commission. *Research in Law and Economics* 14: 191–216.

Brown, Marie. 2017. *Last Line of Defence: Compliance, Monitoring and Enforcement of New Zealand's Environmental Law*. Auckland: Environmental Defence Society.

Cabinet Office (UK). 2006. Regulatory Justice: Sanctioning in a Post-Hampton World.

Crown Law Office. 2013. Solicitor-General's Prosecution Guidelines.

de Prez, Paula. 2000. Excuses, Excuses: The Ritual Trivialisation of Environmental Prosecutions. *Journal of Environmental Law* 12 (1): 65–77.

Environment Canterbury. 2010. Compliance Monitoring and Enforcement Guidelines 2010.

Environment Southland. 2017a. Compliance Policies.

Environment Southland. 2017b. Environmental Compliance Monitoring Report 2016/17.

Forest & Bird. 2018. Cleaning Up: Fixing Compliance, Monitoring and Enforcement in the Dairy Sector.

Gunningham, Neil, and Cameron Holley. 2016. Next-Generation Environmental Regulation: Law, Regulation, and Governance. *Annual Review of Law and Social Science* 12 (1): 273–293. https://doi.org/10.1146/annurev-lawsocsci-110615-084651.

Hall, Matthew. 2014. The Roles and Use of Law in Green Criminology. *International Journal for Crime, Justice and Social Democracy* 3 (2): 96–109. https://doi.org/10.5204/ijcjsd.v3i2.176.

Hampton, Philip. 2005. *Reducing Administrative Burdens: Effective Inspection and Enforcement*. HM Treasury (UK).
Hawkins, Keith. 1984. *Environment and Enforcement: Regulation and the Social Definition of Pollution.* Oxford: Clarendon Press.
Hawkins, Keith. 2002. *Law as Last Resort: Prosecution Decision-Making in a Regulatory Agency*. Oxford: Oxford University Press.
Jurgeleit, Chris. 1996. Insurance Against Liability to Pay Statutory Fines and Penalties. *Victoria University of Wellington Law Review* 26: 735–750.
Marlborough District Council. 2018. Enforcement Policy.
Ministry for the Environment. 2002. A Study into the Use of Prosecutions Under the Resource Management Act 1991: 1 October 1991–30 June 2001.
Ministry for the Environment. 2013. A Study into the Use of Prosecutions Under the Resource Management Act 1991: 1 July 2008–30 September 2012.
Ministry for the Environment. 2016. Compliance, Monitoring and Enforcement by Local Authorities Under the Resource Management Act 1991.
Ministry for the Environment. 2018a. Best Practice Guidelines for Compliance, Monitoring and Enforcement under the Resource Management Act 1991.
Ministry for the Environment. 2018b. Draft Best Practice Guidelines for Compliance, Monitoring and Enforcement under the Resource Management Act 1991.
Naysnerski, Wendy, and Tom Tietenberg. 1992. Private Enforcement. In *Innovation in Environmental Policy: Economic and Legal Aspects of Recent Developments in Environmental Enforcement and Liability*, ed. Tom Tietenberg, 109–138. Aldershot: EE Elgar.
Office of the Auditor-General. 2011. Managing Freshwater Quality: Challenges for Regional Councils.
Office of the Auditor-General. 2019. Managing Freshwater Quality: Challenges and Opportunities.
Resource Management Law Reform Core Group. 1988. Enforcement and Compliance Issues in Resource Management: Ministry for the Environment. Working Paper 30.
Roberts, Julian. 2003. Sentencing Reform in New Zealand: An Analysis of the Sentencing Act 2002. *Australian & New Zealand Journal of Criminology* 36 (3): 249–271. https://doi.org/10.1375/acri.36.3.249.
Sanders, Andrew, and Richard Young. 2007. *Criminal Justice*, 3rd ed. Oxford: Oxford University Press.

Sharpe, Marty. 2017. Councils Issue Nearly Twice as Many Pollution Notices—And Worst Dairy Farms Are Still to Pay Fines. *Sunday Star-Times*, 17 September.

Sharpe, Marty. 2018. Fine of $225,000 for Dirty Dairying Will Go Unpaid Because Companies Are Broke. *Stuff*, 15 May.

Smallman, Elton. 2018. High-Risk Farms the Focus as Aerial Monitoring Looks to Make a Return. *Stuff*, 25 June.

Taranaki Regional Council. 2017. Enforcement Provisions and Procedure under Resource Management Act.

Von Hirsch, Andreas, Andrew Ashworth, and Julian Roberts. 2009. *Principled Sentencing: Readings on Theory and Policy*, 3rd ed. Portland: Hart Publishing.

Waikato Regional Council. 2016. Enforcement Policy.

West Coast Regional Council. 2013. Enforcement Policy.

West Coast Regional Council. 2017. Minutes of the Meeting of the Resource Management Committee.

White, Rob. 2017. Reparative Justice, Environmental Crime and Penalties for the Powerful. *Crime, Law and Social Change* 67 (2): 117–132. https://doi.org/10.1007/s10611-016-9635-5.

Yeung, Karen. 1999. Quantifying Regulatory Penalties: Australian Competition Law Penalties in Perspective. *Melbourne University Law Review* 23 (2): 440–475.

Yeung, Karen. 2004. *Securing Compliance: A Principled Approach*. Oxford: Hart Publishing.

6

Are the RMA's Offences Being Used Appropriately?

The conclusion from the previous chapter was that local authorities prosecute sparingly, for cost, recovery and political palatability reasons, and, when they do prosecute, they seek compensatory outcomes. As noted in Chapter 3, the position taken in this book is that local authorities have been committed to using the offences appropriately, which means that it is necessary to ask if this way of acting, which can be termed the "compensatory gloss" approach, is consistent with criminal law values such as equality of treatment and fairness. This is the task of this chapter.

The compensatory gloss approach implies that local authorities will prioritise two things when prosecuting or deciding whether to prosecute. The first is that any damage caused by the defendant be remedied (or the issue that led to the non-compliance be fixed), and the second is that the council recovers as much of its costs incurred as it can. The first of these things can, generally, be achieved by way of a "compliance-focussed" enforcement order on sentencing, but the second is more problematic. Although the courts seem to agree that fines (90% of which councils receive due to section 342 of the RMA) should usually be the *type* of sentence imposed (as discussed in Chapter 5), the *quantum* of the fine they impose cannot, as a matter of law, be related to the costs incurred

M. Wright, *Responding to Environmental Crimes*, Palgrave Studies in Green Criminology, https://doi.org/10.1007/978-3-030-89250-0_6

by a local authority (*Bay of Plenty Regional Council v Merrie* [2018] NZDC 1621 at [77]). What this means is that, when making prosecution decisions, local authorities will want to keep costs as low as they can and have the offender pay as much of these costs as possible, either by maximising fines or by way of direct payment to the council. In the following sections, the implications of this are considered in relation to three prosecution decisions that a local authority must make—what types of cases to pursue, how to resolve cases "in court" and how to resolve cases "out of court".

Choosing What Types of Cases to Prosecute

The first consequence of the compensatory gloss approach is that prosecutions for breaches of section 9 of the Act (restriction on the use of land) are rare. For instance, the fourth and final prosecution report published by the Ministry for the Environment recorded that only 10% of cases between 1 July 2008 and 30 September 2012 included a section 9 breach, whereas 82% of cases included a breach of sections 12 to 15 breach (restrictions on the use of coastal marine area, certain uses of beds of lakes and rivers, water and discharges of contaminants into environment) (2013, 13). However, this report does not compare the number of prosecutions for each section to the total number of breaches of each section, or the number of enforcement actions for each section, which leaves open the possibility that there were simply far more breaches of sections 12 to 15 than section 9. As such, it is necessary to look to the National Monitoring System, which records the number of enforcement actions for each section (but still not the number of breaches of each section). Using 2016/2017 as an example, 18% of prosecutions brought were for a breach of section 9, whereas 79% of prosecutions were for sections 12 to 15 breaches. However, 37% of the formal enforcement actions were for section 9 breaches, compared with 58% of formal enforcement action for sections 12 to 15 breaches, suggesting that section 9 breaches are more likely to be dealt with by lower level measures than sections 12 to 15 breaches.

This can also be seen in the relative prosecution rates between territorial authorities and regional councils, since the former can only prosecute section 9 (and section 11) cases whereas regional councils can prosecute the full range of breaches. In 2014/2015 territorial authorities commenced, as between them and regional councils, only 26% of the prosecutions (17 as opposed to 48), yet issued 38% of the infringement notices (344 as opposed to 570) and 35% of the abatement notices (399 as opposed to 733). The situation is even more stark in 2015/2016, with territorial authorities commencing, as between them and regional councils, 20% of the prosecutions (11 as opposed to 43), against a backdrop of issuing 51% of the infringement notices (424 as opposed to 402) and 58% of the abatement notices (902 as opposed to 649). Finally, in 2016/2017 territorial authorities commenced, as between them and regional councils, 12% of the prosecutions (6 as opposed to 46), compared with 36% of the infringement notices (355 as opposed to 634) and 17% of the abatement notices (181 as opposed to 1,054).

There are three (inter-related) reasons why the compensatory gloss approach is responsible for this dearth of section 9 prosecutions. First, while these breaches may well have been committed intentionally, they often do not result in any identifiable damage, and therefore there is nothing to be repaired. For instance, the breach could relate to the density provisions in a District Plan, with the offender using their property for guest accommodation when this is not allowed. While this use may have an impact on the neighbouring properties, and threatens the "integrity of the Plan", there is nothing physical to be fixed. Second, section 9 charges can be hard to prove, which means they are likely to be expensive. The starting point is that the default position for section 9 is that you can use land in any way that is not prohibited (by a national environmental standard, regional rule or district rule (as the case may be), or that contravenes sections 176, 178, 193 and 194), and then a series of allowances are also made (for instance, if the use is expressly allowed by a resource consent). Although "use" is defined broadly—for instance, it includes "alter, demolish, erect, extend, place, reconstruct, remove, or use a structure or part of a structure in, on, under, or over land" (section 2(1))—proving a breach of section 9 will often be complex, as the national environmental standard, or regional or district plan, may not be written with the clarity typically found in primary legislation. This means that extensive (and costly) legal advice may be required, along with expert evidence, and the defendant may be more likely to challenge the prosecution case. Third, fines are frequently at the lower end of the spectrum and so the local authority will not recover much of its outlay (via the 90% rule in section 342(1) of the RMA). There is no tariff, or recent official data collected on this, but an analysis of the sentencing decisions for the 2016/2017 financial year shows that fines for section 9 cases were typically around the $20,000 level, with the largest fine during the year being $40,000. It is, therefore, likely that, in the usual course, cost will outweigh recovery.

This can be contrasted with prosecutions for breaches of sections 12 to 15 of the RMA, and especially section 15 (the discharge of contaminants into the environment), the most prevalent type of prosecution. In such cases, there is more likely to be identifiable damage or, as noted in the previous chapter, the courts will make an allowance for harm (*Thurston v Manawatu-Wanganui Regional Council* HC Palmerston North CRI-2009-454-24, 25 & 27, 27 August 2010 at [41]). Further, the default position is that the listed actions are not allowed—by way of example, section 15 begins "[n]o person may discharge…"—making them easier to prove. Finally, average fines for breaches of sections 12 to 15 tend to be much higher than those for section 9. For instance, in the 2016/2017 financial year average fines were over $40,000, with penalties as high as $135,000 imposed. These breaches are, accordingly, much more cost-effective to prosecute—indeed, it is not inconceivable that a local authority could get so accustomed to prosecuting section 15 breaches, in particular, that (90% of) the fines imposed exceeded the cost incurred.

It is important to be clear that the argument is not that a local authority necessarily compares two prosecutable files, one with, say, a section 15 breach and one with a section 9 contravention, and prefers the former because the damage is more apparent and it will be easier and provide a higher fine. Rather, it starts earlier than that. When following up a section 9 complaint or undertaking monitoring (of a consent or generally) the enforcement officer will focus on whether there is any obvious damage. If there is none, then, even if the breach is clearly intentional, he or she will be aware that it could be a long, expensive prosecution, with little in the way of return, and so be more inclined to resolve the matter using lower level measures. As such, many section 9 breaches will not even reach the stage of being considered for prosecution.

However, even when there has been repeated non-compliance and it comes to considering a section 9 breach for prosecution, local authorities are reluctant to bring charges, except as a "last resort". This is because there still may not, despite all the non-compliance, be any identifiable damage and costs will have escalated throughout the process (and a failure to comply to date indicates that the offender may try to defend a prosecution), but the fine will not increase at nearly the same rate. For instance, in *Tauranga City Council v Jacko Basil Holdings Ltd* [2017] NZDC 3273 issues about the property in question being used as two independent dwelling units, in breach of the District Plan, had started back in 2008, with an abatement notice being issued in 2010. After a period when it appeared matters had been resolved, problems started again in 2014, and these continued until charges were eventually laid in 2016. The defendants then tried to challenge the prosecution on the basis that it was outside the limitation period for filing charges (see *Tauranga City Council v Jacko Basil Holdings Ltd* [2016] NZDC 17674). When this was unsuccessful, two of the (initially) three defendants pleaded guilty (charges against the third were, presumably, withdrawn, although this is not mentioned in the sentencing decision). Despite it being intentional offending and involving obstruction of the Council, the end result was only a fine of $23,000 for the company and, upon payment of a $5,000 contribution towards the Council's costs, a discharge without conviction for the individual (a shareholder of the company).

On the other hand, local authorities are willing to move straight from non-compliance to prosecution for unintentional breaches of other sections of the Act, so long as there has been identifiable damage, as they know the court will treat the matter more seriously. For instance, in *Southland Regional Council v Te Wae Wae Dairies Ltd* [2017] NZDC 11466 the Council, responding to a complaint of dairy farm effluent discharging into a waterway, in fact found three discharges (in total) from separate (but neighbouring) farms operated by the defendant. A prosecution was brought, despite it being "an unintentional but careless discharge with no more than a moderate proven effect on the environment" (at [13]). It was not even an ongoing issue—the system had been checked by the Council only six days previously, so the Court proceeded on the basis that it was a "one-off" (at [15])—but was instead caused by the operations manager who was responsible for the system not undertaking checks of the irrigators or disposal fields required by the defendant's operating manuals and procedures. Yet the Court still found the offending to be at the upper end of the moderately serious band and "perilously close to the most serious band" (at [19]), and a total fine of $52,500 was imposed.

The reluctance to prosecute section 9 breaches is a problem because different activities are (primarily) regulated by different sections of the Act. For instance, matters relating to the use of residential and commercial properties will typically be subject mainly to section 9 restrictions, whereas the predominant section that (say) dairy farmers must comply with is section 15. As such, if a residential owner or commercial business and a dairy farmer commit breaches of the same gravity, but the former only receives a low level or informal enforcement response while the latter is prosecuted, then this is unequal treatment.

The obvious response is that the relative prosecution rates of section 9 breaches and breaches of sections 12 to 15 may simply be coming from the application of the *Solicitor-General's Prosecution Guidelines* (Crown Law Office 2013). While not mandatory, as noted in the previous chapter these are typically followed by local authorities, and they set out a test for prosecution that:

> 5.1 Prosecutions ought to be initiated or continued only where the prosecutor is satisfied that the Test for Prosecution is met. The Test for Prosecution is met if:
>
> 5.1.1 The evidence which can be adduced in Court is sufficient to provide a reasonable prospect of conviction – the Evidential Test; and
>
> 5.1.2 Prosecution is required in the public interest – the Public Interest Test.

Choosing what types of cases to prosecute sits squarely within the Public Interest Test. In terms of the factors in favour of prosecution, it might be said that the anticipated penalty for a section 12 to 15 case (one limb of the assessment of seriousness) is typically higher (Crown Law Office 2013, 5.8.1), the offending (especially in the case of dairy farm effluent discharges) is prevalent, and that there is a serious risk of harm (Crown Law Office 2013, 5.8). On the other hand, factors against section 9 cases may include that it will result in a "very small or nominal penalty", the harm was minor, the offence was "not on any test of a serious nature", and other "proper alternatives to prosecution are available" (Crown Law Office 2013, 5.9).

However, many of the arguments in favour of section 12 to 15 prosecutions could be made about section 9 cases. The first point to note is that the maximum penalty for the different breaches (the other limb of seriousness) is the same, and it is arguable that one of the reasons the penalties for such breaches are low is because cases are brought so infrequently. For instance, penalties for dairy farm effluent discharges continue to be raised because cases keep coming before the courts (see, for example, *Southland Regional Council v Baird* [2018] NZDC 11941 at [25]), and so if local authorities instituted a policy of readily prosecuting section 9 cases then it could be expected that there would be a similar jump. Second, as can be seen by the number of formal enforcement responses (and there is likely to be a lot of informal enforcement), section 9 breaches are clearly prevalent. Third, there is still harm caused by many of these breaches, it is just more indirect, and an "allowance for harm" could be made in just the same way it is in discharge cases where no direct damage can be identified. In many ways, section 9 breaches are more serious, as they are often committed intentionally and if a local authority does not maintain the integrity of its plan then the wider community will be led to believe that it can offend with impunity. Such points would also negate the factors against section 9 prosecutions, except that "proper alternatives to prosecution are available", but this factor could equally be said to be true for sections 12 to 15 breaches.

The point is not to suggest that breaches of sections 12 to 15 should be prosecuted less. Rather, section 9 breaches of equal (or greater) seriousness should be treated in the same way as these other breaches, even if they are more expensive to prove and the return is not as high. Otherwise, a defendant who has been prosecuted for an accidental discharge of contaminants could feel a real sense of unfairness when they see someone else, in another industry, intentionally breaching the Act and repeatedly being warned. Further, the public is given the inaccurate message as to the pattern of offending that is happening in their region. Some sympathy could perhaps be had for regional councils, as with restricted enforcement budgets not every prosecution is able to be taken, and so it is understandable they prefer those with the most visibility (in terms of damage) and the greatest recovery. But for territorial authorities there is no such excuse. They do not, by and large, need to compare a section 9 breach with breaches of the other sections of the Act, and so it is difficult to accept that, of all the (likely thousands of) breaches of section 9 that occurred in 2016/2017, only six of these were serious enough for territorial authorities to prosecute.

Local authorities have been able to avoid section 9 cases because the internal and external checks on which prosecutions they are, and more importantly are not, bringing are insufficient. In terms of internal checks, this remains the case despite most (at least large) local authorities now using an enforcement panel for determining whether or not to bring a prosecution. By way of example, Environment Southland's Enforcement Policy notes:

> If a matter is either complex, has a high public profile, requires specific guidance or there is no precedent, then the matter is elevated to an enforcement panel which is convened to make the decision to advance the matter for a legal opinion.
>
> The enforcement panel is comprised of three Environment Southland managers. The panel's main task is to assess the public interest or risk test.
>
> Once a decision has been reached the matter is either re-evaluated or subjected to independent legal review.
>
> …
>
> Once the matter has a legal decision, the entire file is reviewed by the chief executive who makes the ultimate decision to proceed. (2017, 18)

What can be observed first is that for the matter to even be considered for prosecution, the breach must be put forward to the panel for consideration. However, the enforcement officer may instead simply deal with the breach using informal or low-level measures and, as discussed in Chapter 5, can typically do so autonomously. But even if the breach reaches the panel, the process then followed is problematic. As mentioned above, the *Solicitor-General's Prosecution Guidelines* set out a two-stage test, involving first a consideration of whether the "evidence which can be adduced in Court is sufficient to provide a reasonable prospect of conviction" and then, if so, whether "[p]rosecution is required in the public interest" (Crown Law Office 2013, 5.1–5.2). However, as the Environment Southland policy makes clear, the local authority is primarily assessing the second factor, but is doing so before the first factor has been considered by its legal advisors. As such, if the Council's legal advisors subsequently consider that there is insufficient evidence to prosecute, then the enforcement panel's public interest consideration has been pointless. But if there is sufficient evidence to prosecute, then the legal advisors are unlikely to tell the local authority, which is the "expert" for the purpose of the public interest part of the test and has already considered that this is met, that it is not in fact in the public interest. What this means is that the primary "internal" check on a local authority, therefore, is a legal one (whether or not there is sufficient evidence), rather than the more important check on whether it is using its discretion to prosecute (or not) properly.

In any event, whether or not the *Solicitor-General's Prosecution Guidelines* have been followed by local authorities is somewhat academic, as there has, at least historically, been no external checks on council prosecution decisions. Local authorities do not come within the general oversight of the Solicitor-General because the prosecutions they bring are not "public prosecutions" as defined in section 5 of the Criminal Procedure Act 2011 (a point returned to in the next chapter). Further, the Ministry for the Environment does not appear to be involved in reviewing these decisions. Finally, unless there is any public reporting by the local authority that it is *not* prosecuting, which as noted in the previous chapter is rare, the community equally has no idea that serious cases are being dealt with by lower level measures.

There is the potential for this to change. In 2018, an RMA enforcement oversight unit was established in the Environmental Protection Authority (EPA), a Crown Agent established under the Environmental Protection Authority Act 2011, with the aim of improving "the consistency, effectiveness and transparency of council enforcement of RMA rules and decisions" (Parker 2018). Then, in 2020, the EPA was given RMA enforcement powers, including that it may "if satisfied that the performance of the [enforcement] function is necessary or desirable to promote the purpose of [the RMA]", among other things intervene in an enforcement action of a local authority by taking it over and taking any subsequent action (Resource Management Amendment Act 2020 section 96, RMA section 343F(c)). However, the EPA will not be able to intervene in relation to an enforcement action that a local authority has already executed in respect of a person (RMA section 343G(2)), which means that there will be little the Authority can do if the council has already used a lower level response to offending. This would be a major problem, as a common complaint is often not that a local authority has taken no action (in which case the EPA could still intervene, so long as it is within the limitation period), but that the local authority has been too lenient (for instance, issuing an infringement notice when the matter warranted prosecution). Further, the EPA will need to be aware that nothing has been done in response to an incident in order to be able to take enforcement action or intervene and, while it can require information from a local authority (RMA, section 343J), this provision will be of little use if it does not know about the incident in the first place. This is of critical importance because, as has been noted in the previous chapter, reporting on breaches of the Act and informal action that has been taken is typically infrequent.

Resolving Cases in Court

The compensatory gloss approach also impacts on how cases are resolved "in court", as it encourages local authorities to "plea bargain" cases so that they are concluded by way of a guilty plea (or pleas), an agreed summary of facts and a non-natural person defendant. These three resolutions are discussed in turn, but it is first necessary to say something about the term "plea bargain", as this can in fact be divided into a number of different types, including "fact bargaining" and "charge bargaining" (Campbell et al. 2019, 321–333; Duff et al. 2007, 169–180), and "defendant bargaining". Fact bargaining and charge bargaining involve, respectively, the defendant pleading guilty to the charges as alleged in return for the prosecutor agreeing to amend the summary of facts to read more favourably to him or her, and the defendant pleading guilty to one or more charges in return for one or more other charges being withdrawn. Defendant bargaining is not expressly mentioned in the literature (it is dealt with under charge bargaining), but it involves the prosecutor accepting a guilty plea (or pleas) by a defendant (or defendants) and in return withdrawing the charge(s) against a co-defendant (or co-defendants). The more precise terms "fact bargaining", "charge bargaining" and "defendant bargaining" are used in this section as much as possible.

In terms of resolutions by guilty pleas, the final prosecution study notes that in 85% of cases a guilty plea was entered, with this figure ranging between 80 and 91% in the preceding three periods (Ministry for the Environment 2013, 16). While there is no data on the percentage of guilty pleas entered since then, there is nothing to suggest that this has decreased—for instance, in the 2016/2017 sentencing cases reviewed guilty pleas were entered in 33 of the 36 cases (92%). While the high proportion of guilty pleas reflects, to an extent, that local authorities tend to only bring clear cases (as discussed in Chapter 5), it is suggested that a major factor driving this is also charge bargaining. This can be seen by the fact that the outcome of a large number of RMA charges is recorded by the Ministry of Justice as "Not proved", with such definition including where charges are withdrawn by the prosecutor or no evidence has been offered (Ministry of Justice, Letter to author, 23 February 2018). For instance, in 2016/2017 of the 387 charges resolved, only 168 (43%) ended in "Convicted" or "Other proved" (defined as "Youth Court proved, discharge without convictions and Adult diversion/Youth Court discharge"), the remaining 219 (57%) being "Not proved". Given how few cases are argued at trial (and so there will be only a minimal number of acquittals), the implication is that so many charges are "Not proved" because they have been withdrawn by the prosecutor in return for the defendant pleading guilty to other charges.

Guilty pleas are important because of the financial implications to the local authority of not guilty pleas. With not guilty pleas where the defendant elects trial by Judge alone, the cost to the council increases markedly from what would be incurred when a defendant pleads guilty—there are additional administrative hearings and sometimes pre-trial hearings, the council will need to prepare evidence (which may involve the engagement of experts, if this has not already happened), and then conduct a trial (including the possibility of submissions on liability after the hearing has concluded). In terms of recovery, the fine is, at worst, reduced to zero (if the defendant is acquitted) and, at best, only "increases" by the amount of guilty plea discount the defendant would have received. Put another way, the cost of prosecuting may rise from the low tens of thousands of dollars to the hundreds of thousands of dollars (Ministry for the Environment 2016, 25), but, other than in very serious cases, the "increase" in fine will pale in comparison to the additional expense (not to mention the time and inconvenience) to the council. This means that the recovery the local authority receives, as a proportion of its costs, will decrease markedly.

The issue is slightly different when a defendant pleads not guilty and elects trial by jury. The cost difficulty does not arise, as the local authority is no longer required to pay legal expenses once the matter becomes a Crown prosecution and is taken over by the local Crown Solicitor's office (although some councils may still pay for the preparation of evidence (Bay of Plenty Regional Council 2017, 54)). However, local authorities believe that there "have been a number of perverse jury acquittals for regulatory offences of this nature" (Local Government New Zealand 2015, Appendix A), and there will obviously be no recovery if the offender is found not guilty (the local authority still receives 90% of the penalty even if the defendant elects trial by jury). For instance, in *Bay of Plenty Regional Council v Rotorua District Council* DC Rotorua CRI-2017-063-3437, 27 March 2018, the prosecuting regional council was concerned that the defendant territorial authority may be acquitted if the jury became aware that they will (as ratepayers) be responsible for any fine, and the Regional Council applied successfully on this basis to have the trial transferred from Rotorua to Hamilton. An appeal against this decision was allowed (*Waste Management NZ Ltd v Bay of Plenty Regional Council* [2018] NZCA 171), but the mere fact that the local authority applied to move the trial and its reasoning for doing so is what is critical here.

In terms of agreement on facts, there is no data on the number of disputed facts hearings, but they appear to be relatively rare; indeed, of the 36 cases considered in 2016/2017, none were subject to such a hearing. It is much more likely that one will see reference in a sentencing decision to a disputed facts hearing having been considered, but that matters were resolved by the time of sentencing. The implication is that negotiations pertaining to the summary of facts are rife— and this has in fact been acknowledged by the courts (see *Waikato Regional Council v Fulton Hogan Ltd* [2018] NZDC 2711 at [40])—and these nearly always seem to result in an agreed document. Given that the summary of facts is prepared by the local authority, and so it will state the facts at their most favourable to the prosecution to begin with, it will only be amended "down" to favour the defendant.

Agreement on facts is also important for financial reasons, as a failure to achieve this means that a disputed facts hearing, and the time and cost associated with it (preparation of evidence and legal expenses), will be required. While a disputed facts hearing can also lead to a higher fine, in that the prosecutor may be able to "sheet home" a greater level of harm or culpability, and the discount for a guilty plea can be affected when such a hearing is required (*Hessell v R* [2010] NZSC 135, [2011] 1 NZLR 607 at [61]), it can also lead to a reduced fine if the defendant has been successful in the hearing. It becomes a matter of weighing up the possible increase in the fine, which will depend on the local authority being successful in proving the facts and this translating into an increased fine, against the (certain) cost that will come with arguing the matter. Given how risk averse councils are, as discussed in the previous chapter, it is no surprise that this course is not favoured.

Finally, the preference for non-natural persons can partially be seen in the fact that a (slightly) higher proportion of RMA convictions are against companies than individuals—for instance, of the 59 defendants in the cases reviewed in the 2016/2017 financial year, 32 (54%) were companies and 27 (46%) were individuals. But what is more telling is that, where both a company and an individual could be jointly prosecuted, a company is almost always included in the resolution. It is, of course, the case that sometimes the individual(s) is convicted as well (although the practice relating to this appears to vary throughout the country), but the reverse—an individual is convicted and charges against the company are withdrawn—is incredibly rare, typically happening only when the company has become insolvent before the charges have been resolved.

Obtaining a guilty plea from a company is frequently achieved by the local authority entering into an arrangement with the defendants who have been charged whereby the company will plead guilty in return for the proceedings against the individuals, often the directors, being withdrawn. This course is often not publicised, and is only sometimes referred to in the sentencing decision (see, for instance, *Northland Regional Council v Beejay Stud Ltd* DC Whangarei CRI-2016-088-759-760, 26 October 2016 and *Northland Regional Council v Clear Ridge Station Ltd* DC Whangarei CRI-2016-088-403, 26 October 2016, as referred to in the previous chapter). As such, it is perhaps best seen by the fact that, in 2016/2017, of the 145 separate defendants who had charges resolved against them, charges were only proved against 76 of these defendants (52%) (Ministry of Justice, Letter to author, 23 February 2018). The remaining 69 defendants (48%) had all of the charges against them "not proved"—that is, they were acquitted or, more likely given how few cases are taken to trial, had their charges withdrawn or dismissed following a plea arrangement.

The preference for proceeding against companies arises because the maximum monetary penalty is twice as high for non-natural persons as for natural persons, with such difference sometimes being reflected (albeit not to the same extent) in starting points adopted, and therefore the fine imposed. For instance, in *R v Yealands* [2018] NZDC 4115 the sentencing Judge, after noting that the starting point in a similar, but more serious, case had been $45,000, commented that a "starting point of $15,000 is appropriate in this case, bearing in mind the maximum penalty for Mr Yealands is half that which applied for Babich as a corporate defendant" (at [88]). Although the High Court questioned this position in *Trent v Canterbury Regional Council* [2020] NZHC 767, the Judge observing that "the difference in the maximum monetary penalties does not require a court to necessarily arrive at a monetary starting point sentence which reflects the difference in maximum monetary penalties to which an individual and a company may be exposed" (at [53]), and this was upheld on appeal (*Trent v Canterbury Regional Council* [2021] NZCA 123 at [44]), proceeding against a non-natural person provides the *scope* to impose a higher fine than an individual (which would cap out at $300,000). It also stops the sentencing Judge from moving to a non-monetary sentence if the defendant proves impecuniosity (Sentencing Act 2002, section 15(1)(b)).

Finally, there is little to be gained, in terms of recovery to the local authority, from "related" individuals being convicted along with the company. This is because, as noted in Chapter 4, where the defendants are "one and the same for practical fiscal purposes…the total penalty imposed must be tailored accordingly" (*Kiwi Drilling Company Ltd v R* (1997) 4 ELRNZ 23 (CA) at 28), and this typically means dividing the one penalty between them. A local authority can therefore withdraw the charges against such individuals (often directors/owners) at no "cost" (that is, with no reduction in the fine), in return for the company pleading guilty (thereby achieving a better return for the local authority than a defended hearing).

It is important to be clear that negotiations and resolutions within a prosecution are common throughout the criminal law, and are frequently appropriate. Indeed, as the *Solicitor-General's Prosecution Guidelines* state, "[p]rincipled plea discussions and arrangements have a significant value for the administration of the criminal justice system" (Crown Law Office 2013, 18.11). However, using plea bargaining to achieve the goals of guilty pleas, agreed summaries of facts and non-natural person convictions, on the scale it is happening here, is problematic. This is for a number of reasons.

First, local authorities may be "overcharging" in order to secure the desired resolutions. Overcharging (in its broadest sense) occurs when a prosecutor charges unnecessary defendants and offences, or states the facts higher than it can believes it can prove, so that it has "bargaining chips" to give away during plea discussions. However, this is contrary to the *Solicitor-General's Prosecution Guidelines*, which make the fundamental point that the "nature and number of the charges filed should adequately reflect the criminality of the defendant's conduct as disclosed by the facts to be alleged at trial" (Crown Law Office 2013, 8.1). Further, the *Guidelines* go on to state that the "number or seriousness of charges should not be inflated to increase the likelihood of an offer by the defendant to plead guilty to lesser charges" (Crown Law Office 2013, 8.2), and the same applies to what defendants are prosecuted, the *Guidelines* stating that "[c]harges against multiple defendants should be filed only where that is necessary to put the full picture before the fact-finder, or the person charged has played more than a minor role in the offending" (Crown Law Office 2013, 8.5). This is because overcharging:

> ...means that some charge bargains hold no true advantage for the defence and only for the prosecution. By appearing to reduce the charge(s), the prosecutor obtains a plea of guilty to the offence that should really have been charged in the first place. (Campbell et al. 2019, 324)

There have been recent allegations of overcharging made against local authorities—for instance, in *Gu v Auckland Council* [2017] NZDC 29242 it was argued by the defendants, unsuccessfully, that the charges against them should be stayed because the prosecutor had "simply charged everyone it could identify without proper consideration of whether it had admissible evidence to support charges against all of the defendants", and that this had been "compounded...by the apparent bases on which the prosecutor has withdrawn charges against some of the other defendants" (at [11])—and there is evidence to suggest that it may have happened in the past. For instance, in *Auckland City Council v North Power Ltd* DC Auckland CRN 3004510188 & 191, 27 May 2004 the proceedings began with six defendants (North Power Limited and five of its directors) facing 72 charges, but it was resolved by the defendant company (only) pleading guilty to two of these charges. After setting this out, the Court noted that it (at [21]):

> ... does not wish to discourage in any way the resolution of matters in the criminal jurisdiction by an agreement that an informant will seek leave

> to withdraw less appropriate charges in return for the defendant pleading Guilty to the more appropriate charges. This however does not justify the laying of an excessive number of charges in a "scattershot" manner, including charges against directors where there could be no suggestion of any culpability on their individual parts - e.g. because they were aware of previous infringements and had failed to take steps to prevent a repetition, or had deliberately proceeded to "cut corners" in order to save costs.

Alternatively, local authorities may be securing the desired resolutions by entering into plea arrangements that are unprincipled. This occurs when a local authority has commenced a prosecution strictly in accordance with clause 8.1 of the *Solicitor-General's Prosecution Guidelines*, but then finds it needs to "give away" legitimate defendants, charges or facts in order to get guilty pleas and agreement on facts. But if the charges no longer reflect the "essential criminality of the conduct", and do not "provide sufficient scope for sentencing to reflect that criminality", then this is also contrary to the *Guidelines* (Crown Law Office 2013, 18.16–18.18), because the defendant will be treated more leniently, on sentencing, than they deserve.

Perhaps the best example of this is when local authorities charge both the company and its director(s), and withdraw the charges against the latter in return for the company pleading guilty. For instance, in *Waikato Regional Council v Okawa Ltd* [2018] NZDC 7725 Okawa Ltd and its director, and Greener Earthmoving Ltd and its director, each faced a charge relating to unlawfully permitting the excavation of the bed of a river. The effect of the offending on the environment was "significant" (at [46]) and there was no suggestion that there was insufficient evidence against the directors—indeed, the director of Okawa Ltd was described as having been told of the need for a resource consent and his behaviour to proceed regardless was "reckless", and the director of Greener Earthmoving Ltd was "careless but of the highest degree" (at [47]–[56]). However, at the outset of the sentencing the two companies entered pleas of guilty and the prosecutor sought to withdraw the charges against the two directors.

This is related to the final problem, namely that a local authority might withdraw charges against a more culpable director in return for a guilty plea from the less culpable company if this is what it takes to avoid a trial. The director may well be the reason that the company has not, say, invested in new infrastructure to prevent the systemic problem, yet escapes without a conviction and penalty because there is simply no benefit to the local authority from pursuing him or her as well (given the courts' approach to related parties, as referred to above). This is particularly the case where the director is impecunious, as his or her share of the fine may be reduced (Sentencing Act 2002, section 40(1)) and/or converted to a community-based sentence (Sentencing Act 2002, section 15(1)(b)), which would lower the total expected recovery to the council. It also leads one to recall what Bishop has observed, namely that "solvent offenders are not necessarily the most serious polluters" (2009, 302).

Local authorities have been able to resolve prosecutions by way of plea bargains that involve guilty pleas, agreed summaries of facts and non-natural person convictions because, as with choosing what types of cases to prosecute, there is no oversight by the Solicitor-General or, it would seem, the Ministry for the Environment. Again, these are not "public prosecutions" and the Ministry has not asked questions in its National Monitoring System about the number of charges or defendants in each of the prosecutions brought by local authorities, or their outcomes, focussing solely on the number of cases initiated. This is also not something that the EPA will be able to assist with, as it would appear that its ability to intervene ceases once a charge has been laid (RMA, section 343G(2)).

On the other hand, and unlike the decisions regarding what types of cases local authorities bring, the courts do have some responsibility over these resolutions. While the Judge has no influence on the summary of facts—he or she will likely not see any previous drafts before the agreed summary of facts is filed—charges can only be withdrawn with leave of the Court (Criminal Procedure Act 2011, section 146(1)). As such, the presiding Judge can provide some oversight regarding charge and defendant bargaining. However, other than in extreme cases, the courts will not intervene. In particular, so long as one or more charges are remaining, leave will generally be granted to withdraw the other charges. This may be because even if the Judge refused leave, it is very difficult, practically, to require a local authority to prosecute a charge, as if no evidence is offered the Court will have little option but to dismiss the case.

Nor will the public often have any idea that such plea arrangements have taken place. Only some sentencing decisions make reference to charges that have been withdrawn and the defendants who are no longer part of the prosecution, and local authorities do not publicise this aspect of proceedings. Rather, the press releases issued by local authorities only refer to the end result (and are only issued if the local authority has been successful).

Finally, it is also very difficult to obtain this information from local authorities directly. For instance, a request under the Local Government and Official Information and Meetings Act 1987 to the Otago Regional Council (the local authority with the most prosecutions for 2014/2015, 2015/2016 and 2016/2017) for information pertaining to "[t]he names and number of defendants in each case and the number of charges filed against each defendant" and "[h]ow the charges were resolved against each of these defendants (for instance, guilty pleas before trial, guilty verdicts at trial, withdrawn by prosecutor, acquitted or dismissed by a Judge or jury, or in progress still)" was declined (Otago Regional Council, Letter to author, 24 July 2019). This refusal was pursuant to section 17(f) of the Act, on the basis that a record was not held of this information and so it "cannot be made available without substantial collation or research" (Otago Regional Council, Email to author, 5 September 2019). On the other hand, the Council was able to provide the names of defendant(s) (once convicted), charges for which they were convicted and any penalty. This reinforces the point underlying this section that the main concern of local authorities is the end result of prosecutions, not what needs to happen to get there.

Resolving Cases Out of Court

Third and finally, the compensatory gloss approach impacts on how cases are resolved "out of court", with local authorities sometimes not prosecuting (or discontinuing an extant prosecution) if the defendant has remediated the damage caused and/or compensated the council for its costs. This is typically called "diversion", after the schemes run by the Police (NZ Police 2013), and it can take a variety of forms, ranging from informal to formal. However, regardless of the formality of the process, the idea behind the schemes remains the same—in return for the offender doing one or more things to "make up for" the offending, it will not receive a conviction.

Informal diversion is entirely outside of the court process and, at its simplest, involves the offender fixing the issue it has caused in return for the local authority not taking the matter any further. By its very nature, it is not set out in any enforcement policy, and Brown's survey of local authorities found that, while only four councils replied that they had a "bespoke alternative dispute resolution process", "many officers interviewed said that informal reparation is very much more common than those figures indicate" (2017, 53). In addition to this simple arrangement, another type of informal diversion Brown refers to is "retrospective consents", which "are commonly given to activities already completed and they may also allow for the extension of previously unlawful activities" (2017, 53–54). On both approaches, the council avoids the costs associated with the prosecution process, but achieves the outcome it desired of fixing the issue and/or regularising the position going forward.

Before moving on to formal diversion schemes, there is a "halfway house" option whereby a prosecution is brought and the defendant pleads guilty, but before sentencing the defendant agrees to do certain things—generally, repair any damage and make a contribution to the local authority's costs—and in return the local authority agrees not to oppose the defendant's application in Court for a discharge without conviction. Such arrangements also happen from time to time, as *Bay of Plenty Regional Council v Withington* [2018] NZDC 1800 shows. In this case, after the defendant pleaded guilty a restorative justice conference was held, during which the defendant agreed to apologise, to pay for planting native species in the affected area and to pay half the costs incurred by the Council, and in return the local authority would support his application for a discharge without conviction (at [37]–[38]). While such arrangements do not always work out for the defendant—the sentencing Judge in this case did not consider a discharge without conviction to be an appropriate outcome, and instead convicted and discharged the defendant (because his contribution to the Council's costs was approximately the same as 90% of the fine that would have been imposed)—what can be observed is that the local authority was using an out of court process to get the outcome it was expecting from prosecuting. Put another way, the damage was being remediated and the Council received a contribution to its costs equivalent to what it would have received from (90% of) the fine, but without the risk that the Judge may order a lower level of fine or that the defendant may not pay the fine.

In terms of "formal" diversion, this is practised by various local authorities in New Zealand (Environment Canterbury 2012; Environment Southland 2017; West Coast Regional Council 2013). The schemes typically involve the council commencing a prosecution, but then adjourning it while a set of conditions, such as the remediation of the damage caused, are negotiated with and then completed by the defendant. Unlike the "halfway house" arrangement referred to earlier, if the offender completes these to the satisfaction of the local authority the council will seek leave to withdraw the charges, rather than the defendant having to rely on successfully applying for a discharge without conviction. There is no statutory basis for these arrangements, as section 148 of the Criminal Procedure Act 2011 ("Prosecutor must notify court if defendant completes programme of diversion") only applies to programmes "conducted in relation to any public prosecution", which as noted previously does not include RMA prosecutions; rather, such schemes are entirely a creation of the local authorities.

These arrangements will obviously come at a greater cost to local authorities than informal diversion, in that charges must be filed, disclosure provided and various appearances in court made, but councils recognise this by typically requiring, as a condition, the payment of the costs of not only remediation, but also of the investigation and prosecution. For instance, Environment Canterbury's "Guidelines for implementing Alternative Environmental Justice" note under the heading "5.4 Setting Appropriate Conditions" that:

> As costs and fines are both consequences of a conviction, it is similarly appropriate that the issue of the costs of remediating, investigating

> and prosecuting offences be a part of the [Alternative Environmental Justice] process. Unless the defendant has restricted financial capacity, it is expected that the defendant should make a significant contribution to the costs of remediation, the investigation and/or prosecution. (at 8–9)

The Guidelines then go on, in even more forceful terms, to state:

> *It is appropriate that all costs of the offending be covered by the offender and not the rate-paying community.* This has been referred to by the Courts as a "polluter pays" approach. This is because persons who seek to comply with their duties under the Act are subject to full cost recovery. Therefore *a person who seeks to avoid compliance should similarly be subject to full cost recovery.* It is considered that offenders who have avoided the costs of compliance should pay the Council's reasonable costs associated with the non-compliance, such as the reasonable costs of the investigation and/or the reasonable costs of a prosecution. If costs are not agreed it is Environment Canterbury's discretion to seek costs before a District Court Judge. (at 9, emphasis added)

Similarly, the West Coast Regional Council's policy notes that, under the heading "Eligibility Criteria for Alternative Environmental Justice pathway", one of the criteria is "[a]ll Council investigation costs must be met by the offender" (West Coast Regional Council 2013, 17). The Policy then goes on to state:

> **Meeting the Council's Costs**
> The defendant will be liable for all the costs associated with the process. This includes the investigation and legal costs leading to the decision to undertake Alternative Environmental Justice and the costs associated with the Council attending the Conference (including the independent facilitator costs).
> If one of the parties withdraws from the process then the defendant cannot be held liable for costs. However if this was to occur then Council would incorporate these costs into any sentencing submissions during the court process that follows. (at 18)

What can be observed is that, in a similar vein to the "halfway house" option referred to above, such arrangements remove the risk for the local authority. Rather than the "lottery" as to whether remediation will be undertaken and whether (90% of) the fines imposed will meet the council's costs (and that these will be paid), formal diversion schemes allow the local authority the (relative) certainty of knowing that the remediation costs, and its investigation and prosecution costs, will be covered. Indeed, in *Canterbury Regional Council v Bathurst Coal Ltd* [2019] NZDC 14416, a case where the prosecutor was seeking leave to withdraw all the charges filed on the basis that the defendant had been through its diversionary process, the Judge referred to a report from Restorative Justice Services that "contains a statement from a Regional Council employee recording that the outcome of the Scheme 'typically involves a financial contribution that is three times what a fine would be'" (at [14]).

It is important to recognise that, for the council and the defendant (or putative defendant), there is much to gain from diversionary schemes. In particular, such arrangements may assist with getting any damage repaired promptly and keep costs—for both parties—to a minimum. Further, the Ministry for the Environment has observed that "Environment Canterbury reports that no party who has taken part in Alternative Environmental Justice has reoffended and the feedback on the process has been positive" (2018, 99). However, as noted in the previous chapter, there has been repeat offending in other regions, and as the diversionary scheme did not result in a conviction this was not able to be used to uplift the sentence starting point (*West Coast Regional Council v Johnson Brothers Transport (2006) Ltd* [2018] NZDC 22057).

In any event, from a broader perspective these arrangements can be problematic. First, as Ashworth and Zedner note, the "diversion of cases above a certain level of seriousness should be opposed, insofar as it permits a private negotiation between defendant and state to trump the larger public interest in proportionate penal responses and replaces public prosecution with an individualized, private form of justice or 'contractual governance'" (2008, 26; citing Crawford 2003). This is a major problem with RMA diversion, it is suggested, because low-level measures are already available to local authorities and, as discussed in Chapter 4, are used heavily. As such, the cases considered for formal diversion will already have crossed a "seriousness threshold", which can be contrasted with the Police approach of offering diversion for offences that are "not serious" (NZ Police 2013).

Local authorities therefore try to justify such schemes in another way. For instance, Environment Canterbury's guidelines begin by noting that "[m]any environmental offences do not appear to be the result of deliberate or deceptive activities", but that "significant and permanent consequences can follow from environmental offending, even though it may have been inadvertent", and that it is "with this in mind that [Alternative Environmental Justice] has been developed to fill an identified gap in the 'regulatory toolbox' where an infringement fine does not provide an adequate deterrent, but a prosecution may be overly harsh" (2012, 1). Similarly, the West Coast Regional Council policy states:

> Experience has shown that, in some cases, environmental offending is the result of ignorance of the rules or lack of care rather than outright deceptiveness or deliberate actions. Sometimes in these cases the environmental impact is deemed to be more serious than what would warrant an infringement or abatement notice. However, exposing the offender to the full prosecution process may be too harsh. It is in these 'grey areas' where the Alternative Environmental Justice method could come into play. (2013, 17)

However, the first point to note is that Parliament has explicitly stated that whether or not a person intended to offend is irrelevant to liability (RMA, section 341(1)). Rather, as discussed in Chapter 5, this can be—and is—taken into account at sentencing. But more fundamentally, the local authority has created what is essentially its own sanctioning regime, whereby the local authority "imposes" a more serious penalty than the infringement notice, and even potentially an equal or more serious penalty than arising out of a prosecution (albeit without the conviction), without any protections being provided to defendants (see generally Young 2016; Elias 2017). It is in fact worse than this, as the arrangements technically come within a prosecution, and so appear to have supervision from the courts which may encourage defendants to be part of the scheme, but the Court in fact has no role in setting the conditions or determining if they have been fulfilled. Indeed, the local authorities use the courts more as a threat, than as a protection, as shown by the references in both the guidelines of Environment Canterbury and the West Coast Regional Council to seeking costs from the Court if these cannot be agreed with the defendant (in the case of the former) or if the defendant withdraws from the process (in the case of the latter). This issue is referred to in more detail below, but for now the point is that Parliament has provided local authorities with infringement notices, for less serious offending that requires a punitive response, and criminal offences, for more serious offending that requires a punitive response, and if an intermediate sanction is required then this is surely a matter for Parliament.

Related to this, another reason given by local authorities for having a diversionary scheme is that the "interests of justice may not justify the recording of a conviction against the defendant" (Environment Canterbury 2012, 4). However, there is already a way that guilty defendants can avoid this, by applying to the Court for a discharge without conviction if the consequences of such would be out of all proportion to the gravity of the offence (Sentencing Act 2002, sections 106 and 107). Therefore, the local authority appears to be usurping the role of the courts in making a decision as to whether or not a conviction is required. The matter might be different if a diversionary scheme was used where the defendant accepted responsibility for what happened and agreed to make things right, and it had transpired after filing the charges there was a reasonable doubt as to whether the defendant had committed an offence (this is, essentially, how an enforceable undertaking operates: as to which and the effect these have had, see Pedersen 2019). However, this is not the case, as it is typically a requirement of formal diversionary schemes that the defendant admits the offending (see, for instance, Environment Canterbury 2012, 4).

The next problem with diversionary schemes is that they have the potential to treat people unequally, as not every local authority runs a formal diversionary scheme. As such, two people may commit similar offences in neighbouring areas, yet only one will receive a conviction. This cannot even be put down to regional differences. The local authorities that run such schemes are not saying that they are prioritising the enforcement of certain activities in their areas over others, they are saying that, across-the-board, sometimes prosecutable offending in their region does not deserve a conviction.

The most fundamental problem with arrangements reached under these formal diversionary schemes, however, is that some may be unlawful. In *Osborne v Worksafe New Zealand* [2017] NZSC 175, [2018] 1 NZLR 447, the Supreme Court considered the circumstances in which the payment of a sum of money in return for a prosecutor not commencing, or commencing and withdrawing, a prosecution would amount to an "unlawful bargain". This case arose out of the death of 29 men following explosions at the Pike River coal mine on 19 November 2010. Twelve charges had been laid under the Health and Safety in Employment Act 1992 against a director and chief executive officer of Pike River Coal Ltd. However, Worksafe New Zealand (the central government prosecuting agency) ultimately offered no evidence on the charges after the defendant agreed to make a payment into Court of $3.41 million, such funds to be used in satisfaction of an order for reparation that had been made earlier against Pike River Coal Ltd (and for which no payment had ever been made, given that company had gone into receivership). Family members of two of the men who died in the Pike River disaster applied for judicial review of the decision by Worksafe New Zealand to offer no evidence on, and the decision of the District Court to dismiss, the charges. After being unsuccessful in the High Court and Court of Appeal, they succeeded in the Supreme Court on the Worksafe decision.

The Chief Justice commenced her reasons by stating the principle, not in contention, that "[i]t is contrary to the public interest and unlawful for an arrangement to be made that a prosecution will not be brought or maintained on the condition that a sum of money is paid" (at [1]). The critical question was "whether the Court of Appeal was right to hold that the conditional arrangement made by Mr Whittall to pay the reparations ordered against Pike River Coal was not an agreement to prevent the prosecution but an offer of voluntary payment which WorkSafe was entitled to take into account in making its decision about prosecution" (at [25]). Ultimately, the Court decided unanimously that the "conditional payment was a bargain to stifle the prosecution" (at [101]).

The relevance of this is that, as noted above, it is typically the case that, as a condition of (at least) the formal diversionary schemes, the offender must contribute, in part or whole, towards not only any remediation costs, but also the investigation and prosecution costs of the local authority. Accordingly, where these costs are the main part of the resolution between the council and the defendant, there is a risk that the arrangement will fall foul of the test set out in *Osborne*. Put another way, is the local authority only withdrawing the charges because the offender is paying to remediate the damage and/or reimburse the council for its costs?

It would appear that local authorities are conscious of, and have attempted to insulate themselves from, such a risk, by stating that they are not to benefit from the arrangement. For instance, after discussing factors such as proportionality, achievability/timeframe and rehabilitation, Environment Canterbury's guidelines then state:

> **No Benefit to Environment Canterbury**: other than the recovery of actual and reasonable costs of the remediation, investigation and prosecution (which have already been incurred), no condition set as a part of an AEJ conference may result in either a direct or indirect financial benefit to Environment Canterbury. (Environment Canterbury 2012, 8)

The problem with this is that, in a "normal" case these investigation and prosecution costs would not typically be recoverable, other than to the limited extent provided in the Costs in Criminal Cases Act 1967 and the Costs in Criminal Cases Regulations 1987 (as discussed in Chapter 5). Therefore, the payment of actual and reasonable costs of (at least) investigation and prosecution is much more than a council would generally expect to receive from a prosecution and they are benefitting in this way.

What the "no benefit" condition is impliedly referring to, therefore, is likely to be section 342 of the RMA, which as previously discussed provides that (at least) 90% of the fine imposed is paid to the local authority that commenced the prosecution. This is made clearer in the Environment Canterbury guidelines, after a discussion on other matters such as admission of offending and remorse, in the passages quoted earlier regarding the requirement for the payment of costs. In particular, the statement is made that "[a]s costs and fines are both consequences of a conviction, it is similarly appropriate that the issue of the costs of remediating, investigating and prosecuting offences be a part of the [Alternative Environmental Justice] process", and it culminates with the comment that "[i]t is appropriate that all costs of the offending be covered by the offender and not the rate-paying community" (Environment Canterbury 2012, 8–9).

However, fines in a prosecution are not related to investigation and prosecution costs, and this is for good reason. There is no necessary relationship between the costs incurred by a local authority and the chief determinants of the fine, namely the seriousness of the offence and the culpability of the defendant (Burke 2018). A defendant might be caught "red-handed" intentionally dumping seriously harmful contaminants into a protected waterway and be willing to plead guilty quickly, meaning the local authority should incur relatively little expense. Alternatively, a defendant might commit a more minor breach of a District Plan and go to great lengths to obstruct a prosecution, putting the local authority to huge cost. The diversionary arrangements, therefore, may well be of great benefit to the prosecutor, as they remove all of the risk that (90% of) the fine will not cover the local authority's costs.

Compounding it all is the veiled threat, in the final sentence of Environment Canterbury's discussion of costs, that if the defendant does not agree to the Council's assessment then it may apply for costs to be awarded by a District Court Judge. The basis on which they would do this is unclear, as section 4(1) of the Costs in Criminal Cases Act 1967 provides only for costs on conviction and it is typically the case with formal diversionary schemes that there is no conviction entered. But even if a conviction of some sort was entered (which would seem to defeat the point of the process), then as discussed in Chapter 5 any costs award is unlikely to be greater than "scale", and therefore much less than "actual" or what the local authority would consider "reasonable". Similarly, it is difficult to see how the West Coast Regional Council could enforce its threat that "[i]f one of the parties withdraws from the process … then Council would incorporate these costs into any sentencing submissions during the court process that follows." As noted earlier, the Courts have stated that the costs incurred by the prosecutor are irrelevant to the size of the fine (*Merrie* at [77]), and it is difficult to see how the withdrawal by the defendant amounts to "special difficulty, complexity, or importance of the case" (Costs in Criminal Cases Act 1967, section 13(3)), justifying an award of costs that is above scale.

Finally, although formal diversionary schemes come within a prosecution, as opposed to being an arrangement directly between the local authority and the offender, this will not prevent them from being declared unlawful. As the Supreme Court in *Osborne* observed, it is "immaterial" that, as part of the arrangement, a Judge has agreed to the prosecutor withdrawing the charges (at [71]). Further, it is unlikely to make any difference that, as highlighted in previous sections of this chapter, RMA prosecutions are not "public prosecutions" like the prosecution was in *Osborne*. This is because the jurisprudence on unlawful bargains arose out of private prosecutions (see *The Bhowanipur Banking Corp Ltd v Dasi* (1941) 74 Calcutta Law Journal 408 (PC)), and, in any event, local authorities are clearly performing a "public" function in prosecuting offenders for breaching the RMA.

The rationale for formal diversionary schemes, and the potential problems, is perhaps best seen by an example. In 2018, the West Coast Regional Council prosecuted Elect Mining Ltd for causing a slip that contaminated a public water supply, which led to residents in the township of Ross being on water restrictions for more than two months after silt and clay dirtied the water and clogged water treatment filters. However, the charges were ultimately withdrawn by the prosecutor after the defendant agreed to pay a total of $105,000, which included $15,000 to the Ross community for use in a community project, $70,000 to the District Council for costs to the water supply system, $14,767 to the regional council for investigation costs and $5,232 to cover Department of Conservation costs. A representative from the Council considered this was a "good outcome for the community" and was quoted as saying that "the payment being made by Elect through this process is likely to outweigh any fine that may have been imposed and will also directly benefit the Ross community" (Carroll 2018). This is undoubtedly the case. However, this was clearly also serious offending, arguably deserving of a public hearing in court and censure by way of a conviction if proven. Further, it is hard to imagine that, had the defendant not been in a position to pay this amount of money, the charges would have been withdrawn. As such, the inescapable conclusion is that the money was paid to "stifle the prosecution", yet rather than being concerned about the arrangement, as Worksafe New Zealand was in *Osborne*, the West Coast Regional Council actively promoted it.

That informal diversion takes place is easily explained. As noted in Chapter 5, for many RMA offences there is no direct human victim, meaning that there is no one that the local authority needs to explain its actions to at the time and the breach is unlikely to be referred to in any compliance, monitoring and enforcement report. Further, as discussed in the first section of this chapter, there is no one auditing the cases that do not go to prosecution, meaning that enforcement officers can quickly and quietly resolve cases (see generally Gill 2002, 542). If the public becomes aware of such resolutions then this is typically by chance, and even when it does get raised it is often far too late for anything practically to be done about it given, as discussed in Chapter 5, the limitation period on commencing a prosecution.

The next category of diversion was the "halfway house", such as where a local authority agrees not to oppose an application for a discharge without conviction in return for the defendant doing certain things. As seen in the *Withington* case, this is sometimes stopped by the courts. However, on other occasions the courts have even tried to facilitate a similar sort of resolution. For instance, in *Auckland Council v Liu* [2016] NZDC 14243 a pre-trial hearing was held in relation to some of the charges brought by the prosecution. After dismissing one set of charges, the Judge stated:

> [68] In relation to the charges under the [Proposed Auckland Unitary Plan (PAUP)] remaining and under the Operative Plan, although I have significant doubts as to whether the earthworks charges can be sustained, no ruling has been sought from the Court. In any event, as I understand it the applicant is currently finalising an application for resource consent, which will include remediation, and it seems to me that I should allow this process to finalise, as well as the PAUP decisions process of the PAUP to be finalised.

> [69] At this stage I agree with Mr Savage that in light of the primary decisions of the Court, the prosecutor should reconsider whether or not the remaining charges can be resolved in an alternative way without the need to hear these prosecutions. In that regard *it is likely that the Court would view a resolution of the issue by the granting of a resource consent and conditions thereto as akin to a restorative justice process in the event that the Council elected to seek leave to withdraw the charges.* (emphasis added)

Although the Judge went on to state that "there may be other factors relating to these remaining charges which have not been considered by the Court at this preliminary basis and may support the matter proceeding to trial" (at [70]), this shows that the courts are willing to facilitate the "settlement" of prosecutions out of court. Further, it only encourages local authorities to do this without involving the courts at all, such as with the "retrospective resource consent" referred to previously.

Formal diversionary schemes have been allowed to proceed, to an extent, because the Ministry for the Environment has not taken a position on them. For instance, in its report on Compliance, Monitoring and Enforcement by local authorities under the Resource Management Act 1991 the Ministry simply refers to the use of alternative resolution schemes without comment (although it does so after a paragraph explaining the benefits in many cases of prosecution beyond the community for which the council is responsible) (2016, 26). Then, in its more recent *Best Practice Guidelines for Compliance, Monitoring and Enforcement under the Resource Management Act 1991*, the Ministry merely records the benefits of such schemes that have been reported to it, without suggesting they are, or are not, appropriate (Ministry for the Environment 2018, 99).

However, the main reason that such schemes have been able to proceed is because they have, historically, been allowed to by the courts. Given these schemes rely on charges being filed, adjournments being granted and leave being given to withdraw the charges, the courts have great oversight powers. Yet, for the most part, there seems to have been minimal intervention (at least recorded publicly in sentencing decisions or minutes), with a number of cases going through the diversionary processes each year, seemingly without question.

There are signs, though, that this may be changing. For instance, in the case of *Canterbury Regional Council v Bathurst Coal Ltd* [2019] NZDC 14416 the prosecutor sought leave to withdraw the charges on the basis that the defendant had participated in a formal diversionary scheme, which had resulted in it agreeing to protect, by fencing and other works, a section of a local stream (at [3]). This work was not remedial, in the sense that it did not relate to the offending, but was estimated to cost $50,000 (at [15]). The presiding Judge first observed that "[t]he court exercises oversight of a prosecutorial decision seeking leave to withdraw to ensure that the prosecutor has followed a principled and robust process" (at [25]). After noting that "there is considerable public interest in prosecuting offending under the Resource Management Act 1991", Her Honour then noted that the prosecutor "appears to give significant weight to the anticipated penalty", and in particular that, due to the value of the work to be undertaken, that no fine would likely result (at [38]). The Judge was not persuaded by this, and declined the application to withdraw the charges as she could not be satisfied that the "decision to seek leave is a principled one having followed a proper process" (at [39]). In concluding, the Judge also expressed her disquiet at the unequal treatment of offenders between regions, with only some councils operating such schemes (at [43]). The defendant subsequently pleaded guilty and applied for a discharge without conviction, which was declined, and it was instead convicted and fined $18,000 (*Canterbury Regional Council v Bathurst Coal Ltd* [2019] NZDC 23872).

Finally, there is limited public involvement in, or awareness of, formal diversionary schemes. In terms of public involvement, as in Police schemes victims are included in the process of determining whether diversion will be offered to the offender (Environment Canterbury 2012, 4). However, as noted above, it is often the case with RMA offending that no person or entity is directly affected. This can be observed by the comment frequently made by sentencing Judges that section 24A of the Sentencing Act 2002, which makes an adjournment for an inquiry into restorative justice mandatory if, among other things, there are "1 or more victims of the offence", is not applicable. Given "Public Opinion/Community Views" are considered to be irrelevant to the decision whether or not to make an offer of diversion (Environment

Canterbury 2012, 5), the local authority is therefore generally able to make the decision to offer diversion largely on its own.

In terms of public awareness, if a formal diversionary arrangement is allowed then there will typically be no Court decision issued. Indeed, unless the local authority reports on what has happened, then the public will not know that it has taken place or even who the defendant was. For instance, while the Otago Regional Council advised, pursuant to a Local Government and Official Information and Meetings Act 1987 request, that it had granted 10 diversions in the financial years 2015 to 2018, it withheld the names of the parties under sections 6(a) and 7(2)(a) of the Act (Otago Regional Council, Email to author, 5 September 2019). Such grounds are (respectively) that the making available of that information would be likely to prejudice the maintenance of the law, including the prevention, investigation and detection of offences, and the right to a fair trial, and withholding of the information is necessary to protect the privacy of natural persons, including that of deceased natural persons. In what way it would likely prejudice the maintenance of the law and why it is necessary to protect the privacy of natural persons was not made clear—had they been dealt with in Court they would only get name suppression in the very limited grounds set out in section 200 of the Criminal Procedure Act 2011—but the point is that local authorities are treating such arrangements like confidential "settlement agreements", rather than a matter of public record or interest.

Conclusion

It has been shown that the compensatory gloss approach has had an impact on what types of cases local authorities prosecute, how they resolve cases "in court" and how they resolve cases "out of court". We have also seen that there are problems, from a criminal law perspective, with the way the councils are acting. Among other things, regulatees are being treated unequally, local authorities are acting unfairly and in an unprincipled manner, and some arrangements are, at best, unsuitable given the seriousness of the offending and, at worst, unlawful.

It is important to be clear that there is no suggestion that local authorities are acting in this way for nefarious reasons. In fact, it is quite the opposite—they act in the way that they consider to be the best for their ratepayers, in the sense of trying to get any issues fixed (the damage repaired and non-compliance rectified) and limiting the financial burden by recovering as much of the costs they incur as they can. But this is interpreting "best" in very narrow, case by case, terms, and is not an appropriate way to use the criminal law.

Stepping back, the conclusion from Chapter 5, that the approach councils take to prosecutions appears distinctly "non-criminal", can now be taken one step further. It can, in fact, be called "civil", this branch of the law being:

> … not primarily concerned with censuring the wrongdoer for morally blameworthy conduct. Its principal social function is typically to impose 'obligations of repair' on the wrongdoer, thus encompassing an obligation to pay compensation or restitution to the victim harmed by the wrongdoing. Although remedies within the civil paradigm may impose significant financial burdens on the wrongdoer, they are payable to the victim, not to the state, primarily to repair damage done, not to condemn the wrongdoer. (Yeung 2004, 250)

This can be seen in features such as choosing cases where an external remedy is needed and where recovery will be the greatest (and preferably following lower level attempts to resolve the dispute first), bringing proceedings against a number of defendants and with multiple claims, and then settling against "deep pockets" on a much more limited basis, and resolving cases out of court when the outcome will be better than would be achieved in court. These are all hallmarks of civil proceedings, not prosecutions.

Local authorities have been able to use the offences in this way because there have been no or inadequate checks on them. RMA cases do not come within the oversight of the Solicitor-General and the Ministry for the Environment has shown little interest in undertaking this role. The EPA, with its recently acquired powers, may assist with this function, but it faces barriers in being able to do so. Finally, the actions of the local authorities have, for the most part, been condoned by the courts, who are reluctant to intervene in the exercise of prosecutorial discretion, and the public is only advised of the ultimate outcomes of cases (if at all), not the process undertaken to reach the results. However, this does not explain why local authorities themselves consider it acceptable to act in this way. This is the topic of the next chapter, the last in the case study of the RMA's offences.

References

Ashworth, Andrew, and Lucia Zedner. 2008. Defending the Criminal Law: Reflections on the Changing Character of Crime, Procedure, and Sanctions. *Criminal Law and Philosophy* 2: 21–51.

Bay of Plenty Regional Council. 2017. 2016/2017 Regulatory Compliance Report.

Bishop, Patrick. 2009. Criminal Law as a Preventative Tool of Environmental Regulation: Compliance Versus Deterrence. *Northern Ireland Legal Quarterly* 60 (3): 279–304.

Brown, Marie. 2017. *Last Line of Defence: Compliance, Monitoring and Enforcement of New Zealand's Environmental Law*. Auckland: Environmental Defence Society.

Burke, Andrew. 2018. Fairness, Justice and Repairing Environmental Harm; Reconciling the Reparative Approach to Sentencing of Environmental Crimes with Sentencing Principles. *Environmental and Planning Law Journal* 35: 529–541.

Campbell, Liz, Andrew Ashworth, and Mike Redmayne. 2019. *The Criminal Process*, 5th ed. Oxford: Oxford University Press.

Crawford, Adam. 2003. 'Contractual Governance' of Deviant Behaviour. *Journal of Law and Society* 30 (4): 479–505. https://doi.org/10.1111/j.1467-6478.2003.00267.x.

Crown Law Office. 2013. Solicitor-General's Prosecution Guidelines.

Duff, R.A., L. Farmer, S. Marshall, and V. Tadros. 2007. *Trial on Trial: Volume 3: Towards a Normative Theory of the Criminal Trial*. Oxford: Hart Publishing.

Elias, Sian. 2017. Managing Criminal Justice. *New Zealand Criminal Law Review* [2017]: 316–329.

Environment Canterbury. 2012. Guidelines for Implementing Alternative Environmental Justice.
Environment Southland. 2017. Compliance Policies.
Gill, Peter. 2002. Policing and Regulation: What Is the Difference? *Social & Legal Studies* 11 (4): 523–546. https://doi.org/10.1177/096466390201100403.
Local Government New Zealand. 2015. RMA Sector Position Paper: Draft for feedback.
Ministry for the Environment. 2013. A Study into the Use of Prosecutions Under the Resource Management Act 1991: 1 July 2008–30 September 2012.
Ministry for the Environment. 2016. Compliance, Monitoring and Enforcement by Local Authorities Under the Resource Management Act 1991.
Ministry for the Environment. 2018. Best Practice Guidelines for Compliance, Monitoring and Enforcement Under the Resource Management Act 1991.
NZ Police. 2013. Police Adult Diversion Scheme.
Parker, David. 2018. Resource Management Act Oversight Unit to Be Established.
Pedersen, Ole W. 2019. Revisiting the Role of Negotiation and Trivialization in Environmental Law Enforcement. *Journal of Law and Society* 46 (1): 29–54. https://doi.org/10.1111/jols.12141.
Carroll, Joanne. 2018. Mining Company Avoids Court with Six Figure Slip Settlement. *Stuff*, 31 October.
West Coast Regional Council. 2013. Enforcement Policy.
Yeung, Karen. 2004. *Securing Compliance: A Principled Approach*. Oxford: Hart Publishing.
Young, Ronald. 2016. The Harkness Henry Lecture: Has New Zealand's Criminal Justice System Been Compromised? *Waikato Law Review* 24: 1–13.

7

Explaining the Inappropriate Use: Form Versus Substance

It was argued, in Chapter 6, that the compensatory gloss approach that local authorities have adopted has led them to use the offences inappropriately, and that they have been allowed to act in this manner due to a lack of oversight or accountability. However, this does not explain why local authorities themselves consider it acceptable to act in this way. This is the topic of this chapter, and the question is considered from a systemic perspective—is the legislation driving local authorities to make prosecution decisions in a manner that is not only contrary to criminal law norms, but in fact how a would-be litigant would approach a civil dispute? To answer this, it is necessary to consider the drafting (and interpretation by the courts) of the RMA's offences, as well as revisiting the penalties imposed and looking in more detail at the institutional arrangements for prosecuting.

M. Wright, *Responding to Environmental Crimes*, Palgrave Studies in Green Criminology, https://doi.org/10.1007/978-3-030-89250-0_7

Expansive Offences and Limited Defences

Before the RMA, there were a number of different statutes that dealt with the management of natural and physical resources in New Zealand, such as the Water and Soil Conservation Act 1967, the Clean Air Act 1972 and the Town and Country Planning Act 1977. However, towards the latter stages of the 1980s it became apparent that there was widespread dissatisfaction with the legislation, and the Government of the day initiated a Resource Management Law Reform process. It was this process, which took four years and cost in excess of $8 million, that ultimately resulted in the RMA (Palmer 2013, 6), and so it will be referred to in some detail throughout this chapter to explain where the offences, penalties and institutional arrangements in the RMA came from.

One of the concerns raised in the reform process was that prosecuting under the legislation existing at the time was difficult (see, for example, Resource Management Law Reform Core Group 1988a, 5), and so the RMA set about addressing this. One way in which it did so was procedurally, which meant stripping away various restrictions in the previous legislation, such as who could prosecute and notification requirements, and adopting a standard timeframe for bringing proceedings. The other way it tried to make prosecution easier, which is the focus of this section, was in the definitions of the offences, and this was achieved by casting the liability net wide, both in terms of who can be liable and for what, and limiting the excuses that can be relied on.

The starting point is section 338(1) itself, which commences by stating that "[e]very person commits an offence against this Act who *contravenes, or permits a contravention of*, any of the following…" (emphasis added). Looking first at "contravenes", which is defined as including "fail to comply with" (section 2(1)), this is broad enough to mean that, along with individuals, companies can be "directly" liable under the Act. For instance, in one of the leading cases on RMA liability, *McKnight v NZ Biogas Industries Ltd* [1994] 2 NZLR 664 (CA), there was no argument that the defendant company, which had been responsible for the installation of the 450,000 litres capacity elastic bladder that had subsequently burst, may have directly contravened the Act; rather, the case related solely to whether it had *caused* the discharge (a matter which is discussed below). Further, when a challenge to direct corporate liability was made, on the basis that the defendant company had contracted out the management of the business to another person or entity and so had not contravened the Act, this was given short shrift by the courts (see *Southland Regional Council v Sandstone Dairy Ltd* DC Invercargill CRN06017500043, 44 & 91, 14 December 2006; appeal against conviction dismissed: *Sandstone Dairy Ltd v Southland Regional Council* HC Invercargill CRI 2007-425-1, 15 May 2007).

However, even if the company is not directly liable, it may still be vicariously liable pursuant to section 340(1)(a) of the RMA. This provision states that "[w]here an offence is committed against this Act by any person acting as the agent (including any contractor) or employee of another person, that other person shall, without prejudice to the liability of the first-mentioned person, be liable under this Act in the same manner and to the same extent as if he, she, or it had personally committed the offence". This has been held to not be a standalone offence, but rather supplements section 338(1) and does not need to be referred to in the charging documents (*Fulton Hogan Ltd v Canterbury Regional Council* [2018] NZDC 17099 [*Fulton Hogan Ltd 2018*]; appeal dismissed: *Fulton Hogan Ltd v Canterbury Regional Council* [2019] NZHC 1767 [*Fulton Hogan Ltd 2019*]). This means that there is scope for a local authority to commence proceedings against a company, instead of or in addition to a responsible individual, leaving it to determine the precise basis for liability at a later date.

The next point is that the defences for those charged vicariously must be proved by the defendant and are quite restricted. If the person charged is a natural person, the defendant must prove that "he or she did not know, and could not reasonably be expected to have known, that the offence was to be or was being committed", or that "he or she took all reasonable steps to prevent the commission of the offence" (section 340(2)(a)(i)). If the person charged is a non-natural person, it must prove that "neither the directors (if any) nor any person involved in the management of the defendant knew, or could reasonably be expected to have known, that the offence was to be or was being committed", or that "the defendant took all reasonable steps to prevent the commission of the offence" (section 340(2)(a)(ii)). In both instances, the defendant must also prove that he, she or it took "all reasonable steps to remedy any effects of the act or omission giving rise to the offence" (section 340(2)(b)).

Further, as noted above, the local authority does not have to specify that it is relying on section 340(1) of the Act in the charging documents, which can make the availability of the defences problematic. For instance, the prosecutor may specify only that the defendant's conduct "contravenes" under section 338(1) in its charging document and ultimately opt against relying on vicarious liability at trial, seemingly precluding the defences from being run. In *Southland Regional Council v Sandstone Dairy Ltd* the Court countered this possibility by stating that "any defendant can make out on the balance of probabilities that he is only liable under section 340(1) and that the defences under section 340(2) apply" (at [35]), but as noted by counsel for the defendant in *Fulton Hogan Ltd 2018*, this would seem "contrary to established principles concerning the burden of proof in criminal proceedings" (at [40]). The preferred course, in *Fulton Hogan Ltd 2018*, was that "the s 340(2) defences apply whenever vicarious liability is imputed simply by operation of s 340(1)" (at [49]), which was affirmed in the High Court as it "serves to give the statutory defences a universal application. This would otherwise be arbitrarily limited not by reference to the substance of the prosecution but by the happenstance of a charging document" (*Fulton Hogan Ltd 2019* at [68]). However, it is not clear how easy this will be to apply in practice, with the availability of the defence still seemingly depending, to a large extent, on the way that the prosecution runs its case at trial.

The third way that a company can be liable is "indirectly", by "permitting a contravention" of the Act. The word "permits" was considered in *Waikato Regional Council v Hillside Ltd* DC Te Awamutu CRI-2008-019-500880-919, 20 July 2009 and given an expansive meaning, with the Judge stating (at [262]):

> I find myself in agreement with the submissions of Mr Pilditch based on several dictionary definitions, and having regard to the broad and important purpose of the Act, that the concept of "permits" encompasses:

a) Providing or affording opportunity for acts or omissions;
b) Allowing acts or omissions to be done or occur;
c) Acquiescence to acts or omissions;
d) Abstinence from prevention of acts or omissions;
e) Tolerating acts or omissions.

The wide interpretation of this term also considerably opens up another group of persons to liability, namely directors or persons involved in the management of companies. For instance, in *Hillside Ltd* itself the defendant company and its three directors faced charges relating to the unlawful discharge of dairy effluent, with the allegation against the three individuals being that they permitted the contravention of section 15 of the Act. It was sufficient to ground liability that they were directors and persons concerned in the management of the business who were aware of, and did not stop, the problems.

This interpretation by the courts is important because the liability of directors is one area where those involved in the Resource Management Law Reform process had urged caution, and where the Government had seemingly followed their advice. A working paper the Resource Management Law Reform Core Group prepared on Enforcement and Compliance Issues in Resource Management had observed that a "suggestion has been made that for deliberate breaches of environmental codes, criminal liability be imposed on directors of companies, (as it the practice in USA and Canada)" and it was commented that "[w]here such behaviour is flagrant and destructive, and clearly of a criminal nature, there may be justification for this approach" (1988a, 44). However, it went on to note that "legislation of this kind would be something of a departure in New Zealand, the implications of which would need to be considered carefully" and that, as "it is a company law matter, and outside the scope of RMLR, it is probably inappropriate to discuss it any further here" (1988a, 44). Accordingly, the mechanism in the RMA for holding directors and management liable, section 340(3), is quite restrictive, providing:

> If a person other than a natural person is convicted of an offence against this Act, a director of the defendant (if any), or a person involved in

> the management of the defendant, is guilty of the same offence if it is proved—
>
> (a) that the act or omission that constituted the offence took place with his or her authority, permission, or consent; and
> (b) that he or she knew, or could reasonably be expected to have known, that the offence was to be or was being committed and failed to take all reasonable steps to prevent or stop it.

Indeed, section 340(3) has been described as "a rather complex offence formula not requiring full *mens rea*, but requiring a degree of actual or constructive culpability or fault, with the onus on the prosecution to prove lack of due diligence" (Grinlinton 1992, 19). The rationale for this wording is unclear, but it would appear that the Government simply imported paragraph (a) from two of the previous resource management statutes (section 47L(3) of the Petroleum Act 1937, as amended by section 3 of the Petroleum Amendment Act 1975, and section 50(3) of the Clean Air Act 1972), but then tried to limit it by replicating the "due diligence" provision found in section 340(2) of the RMA (discussed below).

This is an awkward combination, but because *Hillside Ltd* made it easier for directors to be prosecuted indirectly (for "permitting a contravention"), there is rarely a need to rely on section 340(3) and it is used infrequently. Indeed, in the one of the few cases where it has been successfully used, *Canterbury Regional Council v Foster* DC Christchurch CRI-2013-009-2076, 30 May 2014, the Judge noted that, if section 340(3) had failed, he would have amended the charges to "reflect [his] finding that the evidence presented by the Council also establishes that Mr Foster permitted the offences as identified in or consistent with the *[Hillside Ltd]* decision" (at [30]). Further, local authorities appear to have adjusted their practices to maximise the chances of directors or management coming within the definition of "permitting". For instance, the Waikato Regional Council's *Basic Investigative Skills for Local Government* state that it is "good practice (when issuing an abatement notice to a company) to serve the original of the notice to the company's registered office and also to serve copies of that notice to each of the company directors, as well as any site liaison officer or site environment officer". This is because it "can often bring attention and pressure from the directors onto site management to reach compliance" and "*if there is future non-compliance, it can clearly be shown that the directors have been made aware of previous non-compliance issues*", which "would make it very difficult to plead ignorance or establish a defence under section 340 and 341 RMA" (2016, 104, emphasis added).

To summarise to this point, in addition to individuals, liability has also been extended directly, vicariously and indirectly to companies. Further, while a relatively limited role was originally given to director/management liability, by way of section 340(3), the courts have stepped in to expand this greatly by interpreting "permitting a contravention" broadly. It is no surprise, then, that the RMA has been described as having a "wide net of potential liability for criminal offending" (*Fulton Hogan Ltd 2018* at [55]), with it being entirely conceivable that a breach of the Act could, simultaneously, be committed by an individual, his or her corporate employer and its directors/management.

The next matter to consider is what breaches these people and entities can be prosecuted for. Section 338(1) lists these as contraventions (or permitting the contravention) of:

(a) Sections 9, 11, 12, 13, 14, and 15 (which impose duties and restrictions in relation to land, subdivision, the coastal marine area, the beds of certain rivers and lakes, water, and discharges of contaminants):
(b) any enforcement order:
(c) any abatement notice, other than a notice under section 322(1)(c):
(d) any water shortage direction under section 329.

For present purposes, these breaches can be separated into two groups—the first is breaches of sections 9, 11, 12, 13, 14 and 15 (paragraph (a) above), and the second is breaches of an enforcement order, an abatement notice or any water shortage direction (paragraphs (b)–(d) above). The difference is that the former are "one-step" breaches (in that a duty or restriction is being directly breached) and the latter can be referred to as "two-step" breaches (in that a Court order or an administrative notice or direction must first be made, and it is the breach of this order, notice or direction that is an offence). However, in many instances a two-step breach will also be a one-step breach—for instance, someone may have received an abatement notice ordering them to cease discharging a contaminant, which if breached will be both an offence against section 338(1)(a) and an offence against section 338(1)(c). The focus here is on breaches of section 15 and an abatement notice, as they are the most prevalent prosecutions in each group.

Looking first at section 15 breaches, subsection (1)(b) (by way of example) provides that:

> No person may discharge any … contaminant onto or into land in circumstances which may result in that contaminant (or any other contaminant emanating as a result of natural processes from that contaminant) entering water … unless the discharge is expressly allowed by a national environmental standard or other regulations, a rule in a regional plan as well as a rule in a proposed regional plan for the same region (if there is one), or a resource consent.

It can immediately be observed that this provision is already quite wide, as the default position is that the discharge is prohibited and the discharge of contaminants onto or into land only needs to be in circumstances where it *may* enter water (unless expressly allowed). However, it is made even wider by the interpretation of the words "contaminant" and "discharge". In terms of the former, the definition in section 2(1) is broad, including "any substance…or energy…or heat, that either by itself or in combination with the same, similar, or other substances, energy, or heat" that when discharged into water or onto or into land or into air "changes or is likely to change" the "physical, chemical, or biological condition". As such, there is no room for a consideration as to the extent or effect of the change when considering liability (*Works Infrastructure Ltd v Taranaki Regional Council* (2002) 8 ELRNZ 84 (HC)). In terms of the word "discharge", which includes "emit, deposit, and allow to escape" (section 2(1)), this has been interpreted as extending to "cause to discharge" (*McKnight* at 670). What this means, essentially, is that the test is simply whether there is a causal connection between the person and the discharge. As such, if an employee on a dairy farm fails to move an irrigator, and this results in ponding of effluent or its application to saturated ground, contrary to the relevant regional plan, he or she has caused the discharge and will potentially be liable, along with his or her manager, and their employer, to a conviction and penalty in the tens (or hundreds) of thousands of dollars.

Turning then to a breach of an abatement notice, the offence provision is also quite broad due to two interpretations from the courts. The first interpretation is that abatement notices can be indefinite (*Roberts v Northland Regional Council* [2014] NZHC 284 at [24]–[25]), which means that a local authority can prosecute for breaches that occurred a long time after the notice was originally issued. This is the case even when a later abatement notice has been issued, and complied with, as happened in *Bay of Plenty Regional Council v Hood* [2017] NZDC 25156. While there was said to be "something unfair" about this, with the "sense of unfairness…increased having regard to the general nature of the earlier notice and the specific requirements of the later notice which were met by the defendant", and that "the circumstances here suggest a lack of true connection between the earlier abatement notice and the laying of charges for breach of that notice after compliance with the later abatement notice", there was no dispute that an offence had still been committed (at [28]). The second interpretation is that if the recipient has not appealed the abatement notice and is later prosecuted for breaching it, he or she cannot argue at trial that the notice was wrongly issued (*Waikato Regional Council v Huntly Quarries Ltd* [2004] NZRMA 32 (DC)). The abatement notice is, therefore, a powerful measure that can extend liability greatly, and it is no surprise that the Ministry for the Environment says it should be an enforcement officer's "primary tool" (2018a, 88).

The conduct that can be caught in the aforementioned "wide net of potential liability for criminal offending" is, accordingly, also very broad, and seemingly at odds with the conclusion from the working paper on compliance and enforcement issues in the Resource Management Law Reform process that it:

> …is important that offences be clearly defined. It is suggested that this will partially be achieved by determining exactly what types of conduct it is wished to prohibit and setting out a number of specific offences

> rather than attempting to achieve the same result by means of one very generally worded offence provision. (Resource Management Law Reform Core Group 1988a, 38–39)

Indeed, when one considers the combination of who can be held liable and for what, it is suggested that the words of Farrier, in discussing Australian environmental criminal law in the early 1990s in a chapter entitled "In Search of Real Criminal Law", are more apt for the RMA:

> Little attention has been paid to the details of the definition of offences. The main concern of governments, propelled by the conservation movement, has been that they should be framed in broad terms to ensure that there are no loopholes. (1992, 80)

Turning then to the mental element of the offences, the issues surrounding what *mens rea*—if any—should be included in the criminal offences in the new resource management statute was also raised in the Resource Management Law Reform working paper on enforcement and compliance issues (Resource Management Law Reform Core Group 1988a, 42–44). Under the previous legislation, it was common for offences to be strict liability, although this was not universal. However, some agencies suggested that even this was too hard to prove, and that the offences should be absolute liability, meaning no defence of total absence of fault would be available (Resource Management Law Reform Core Group 1988a, 43–44). On the other hand, it was also suggested that it would be inappropriate for "almost all offences in the environmental sphere to be strict or absolute liability", citing the "fundamental principle of our legal system that any truly criminal offence requires a guilty mind" (Resource Management Law Reform Core Group 1988a, 43–44). It was also commented that if imprisonment was going to be an option there would need to be an explicit mental element (Resource Management Law Reform Core Group 1988a, 46).

Despite these warnings, section 341(1) of the RMA provides that "[i]n any prosecution for an offence of contravening or permitting a contravention of any of sections 9, 11, 12, 13, 14, and 15, it is not necessary to prove that the defendant intended to commit the offence", and imprisonment is available for all offending. It is not clear why section 341(1) does not apply to two-step breaches, but the courts have nevertheless held that these are common law strict liability offences (*Auckland City Council v Selwyn Mews Ltd* DC Auckland CRN2004067301-19, 18 June 2003 at [93]–[100] (breach of an enforcement order); *Waikato Regional Council v Huntly Quarries Ltd* [2004] NZRMA 32 (DC) at [56]–[67] (breach of an abatement notice)).

For those provisions covered by section 341, the two statutory defences in subsection (2) replace the common law defence of "total absence of fault" (*NZ Biogas Industries* at 669). These defences, essentially "reasonable action in a severe emergency or unforeseeable act beyond the defendant's control" (Campbell 2015, [20.45]), must be proven by the defendant and have the added hurdle that the defendant must have "adequately mitigated or remedied" the "effects of the action or event". This has led to the offences being described as "closer to an absolute liability standard than common law strict liability" (Campbell 2015, [20.45]).

Subsection (3) of section 341 requires that the defendant give notice that he or she intends to rely on a subsection (2) defence, and the facts that support his or her reliance, within seven days of service of the summons or such further time as the Court may allow. However, this sits awkwardly with the Criminal Disclosure Act 2008, as the prosecutor may not even have completed "Initial disclosure" within the seven days, which must be completed "not later than the applicable date" and which is generally 15 working days after the commencement of the proceedings (section 12(1)). For this reason (among others), an extension is frequently given, but that does not detract from the wider point that the defendant is being forced to give advance notice of its case if it intends to rely on this defence. No reason was given why this is required and it is unusual for the criminal law.

This all leads to the conclusion that if, as suggested in Chapter 3, the criminal sanction is properly reserved for substantial wrongdoing, assessed predominantly by reference to harm and culpability, then the RMA's offences are failing to meet this test. Harmless conduct is captured alongside the harmful, actors who are only tangentially linked to the conduct are included and, by not requiring the prosecution to prove *mens rea* in any of the breaches, the non-culpable are caught up with the culpable. While there are avenues by which defendants can avoid liability, these are very limited, and by requiring the defendant to prove them, and in some instances notify the prosecution of its intention to do so, they interact awkwardly with fundamental criminal law rights such as the presumption of innocence.

A Focus on Insurable, Compensatory Monetary Penalties

Another goal of the Resource Management Law Reform process was to make prosecuting more worthwhile, and the Government tried to achieve this in the RMA in three (related) ways. First, it adopted the highest maximum penalty from any of the previous resource management statutes ($150,000 from the Water and Soil Conservation Act 1967, which was increased to $200,000 in the RMA during the Parliamentary process), on the basis that this would result in increased fines (Resource Management Bill 1989 (224-1), explanatory note at xi). Indeed, in 2009 when fines were still not considered to be high enough, "[b]ecause the ceiling on fines in New Zealand is relatively low" (Cabinet Paper 2009, 25), the maximum monetary penalty was increased again, to a fine not exceeding $300,000 for natural persons and a fine not exceeding $600,000 for non-natural persons (Resource Management (Simplifying and Streamlining) Amendment Act 2009). As shown in Chapter 4, each of these moves resulted in average fine levels rising, albeit still only to a level that has been described as "modest".

Second, the Government sought to make prosecution more worthwhile by making it explicit that local authorities would receive (at least 90% of) any fines imposed. The relevant provision of the RMA is section 342, which states that:

> (1) Subject to subsection (2), where a person is convicted of an offence under section 338 and the court imposes a fine, the court shall, if the proceedings in relation to the offence were commenced by or on

behalf of a local authority, order that the fine be paid to that local authority.

(2) There shall be deducted from every amount payable to a local authority under subsection (1), a sum equal to 10% thereof, and this sum shall be credited to a Crown Bank Account.

(3) Notwithstanding anything in subsection (2), where any money awarded by a court in respect of any loss or damage is recovered as a fine, and that fine is ordered to be paid to a local authority under subsection (1), no deduction shall be made under subsection (2) in respect of that money.

(4) Subject to subsection (2), an order of the court made under subsection (1) shall be sufficient authority for the Registrar receiving the fine to pay that fine to the local authority entitled to it under the order.

(5) Nothing in section 73 of the Public Finance Act 1989 shall apply to any fine ordered to be paid to any local authority under subsection (1).

The origins of section 342 are unexplained. There was no equivalent provision in any of the previous resource management statutes, and the payment of fines to local authorities was not explicitly discussed in the enforcement and compliance working paper published as part of the Resource Management Law Reform process, although reference was made (at different points) to the cost of prosecution and funding arrangements. In particular, there was support for local authorities being able to "pass on the true cost of enforcement" (Resource Management Law Reform Core Group 1988a, 35), and it was observed that:

> The funding of enforcement is an important question because even if prosecutions are successful there is usually a financial burden carried by the prosecuting authority which in itself can act as a deterrent to bringing prosecutions in other than the clearest of cases. Environmental prosecutions can become exceedingly protracted and complex (ie costly). (Resource Management Law Reform Core Group 1988a, 53)

Curiously, the most direct reference to fines being returned to the prosecutor is made in the Executive Summary to the enforcement and compliance working paper, with the statement that fines "should not be considered a means of raising revenue" (Resource Management Law Reform Core Group 1988a, 6), but this comment is not found (let alone expanded on) in the body of the text. Finally, section 342 is not mentioned in the explanatory note to the Resource Management Bill and does not appear to have been discussed in the Parliamentary debates.

These points, along with the fact that the Minister responsible for introducing the RMA, Sir Geoffrey Palmer QC, has in recent years raised issues with the return of fines to local authorities (Palmer 2016, 48), suggests that it was not necessarily a conscious policy decision to include section 342. It is seemingly based (given the similarity in wording) on the provision it excludes from applying, section 73 of the Public Finance Act 1989, and so perhaps the best explanation is that it was included by the drafters of the Resource Management Bill to rule out any uncertainty as to whether section 73 would apply. Such uncertainty would arise because section 73 is only engaged where the local authority or other organisation is "specifically empowered" to prosecute by an Act, yet, as noted in Chapter 5, the only statutory instructions in the RMA regarding enforcement are the requirements that local authorities "enforce the observance" of their plans and any national environmental standards (RMA, sections 84(1) and 44A(8), respectively).

The inclusion of section 342 is, in one sense, understandable, in that it seems fair that fines are paid to the local authority that instituted the proceeding when it is that party which is responsible for the costs of prosecuting. However, the problem with the provision is that, coupled with the emphasis placed on monetary penalties being imposed (by adopting high maximum penalties, and then raising these), it makes it look, to the local authorities, as though the purpose of the fines is compensation for the investigation and prosecution costs of the local authority, rather than punishment. Yet compensation for these costs is not typically associated with the imposition of penalties, and this is for good reason. As Horder has noted, in response to Tadros's suggestion that "penalties are aimed at ensuring that the costs of breaching regulations are borne by those who breach it rather than those that suffer from the breach" (Tadros 2010, 174):

> Many of the significant costs of implementing regulatory penalties—the employment of officials, outlay in relation to any investigation and the making of findings, together with the costs of fine recovery and of conducting appeal hearings—must in fact commonly be borne by the state rather than by the offender. This is simply because it would be unduly harsh and out of proportion to make the wrongdoer bear all these costs, other than in exceptional cases. So, it is not strictly true to say that penalty schemes aim to ensure that, 'costs of breaching regulations are borne by those who breach it,' nor, in any legal system that aspires to do justice in a fair and proportionate way, should it ever be true. (2014, 120)

The matter is then compounded by the fact that there is no express prohibition in the RMA on insuring against fines, which as noted in Chapter 5 has led the courts to declare that they have no power to declare an insurance policy against these void. It is, perhaps, understandable that insurance was not discussed in the Resource Management Law Reform process, as it was not necessarily a live issue at the time. However, this does not explain why, when other statutes were amended to prohibit such insurance (see, for instance, the changes made to the Health and Safety in Employment Act 1992 in 2002), no changes have been made to the RMA in this regard. Ultimately, if an insurance company is paying the fine on behalf of the defendant then it is hardly a punishment, but rather a further indication that what is most important is that the fine is in fact paid (as without insurance the fine may not be paid or heavily reduced, pursuant to section 40 of the Sentencing Act 2002, if the defendant is impecunious) so that the local authority can recover some of its costs.

The position under the RMA can be contrasted with that for fines imposed in central government regulatory prosecutions. Perhaps the most similar situation to the RMA, in terms of the number of prosecutions brought each year and the penalties involved, is the Health and Safety at Work Act 2015. In prosecutions under this Act, fines are not returned to the prosecutor (except in the broad sense that Worksafe New Zealand is a Crown Entity, and any fines paid in response to a prosecution it has brought are, therefore, property of the Crown), with provision instead made for the Court to "order the offender to pay to the regulator a sum that it thinks just and reasonable towards the costs of the prosecution (including the costs of investigating the offending and any associated costs)" (section 152(1)). Insurance against fines is expressly prohibited (section 29). This means that fines are clearly demarcated as punishment and costs awards are compensation for investigation and prosecution costs, and there is no financial incentive to prefer one prosecution over another, or one type of penalty over another.

This last point about preferring one type of penalty over another is important because, as noted in earlier chapters, fines are not the only penalty that can be imposed for RMA offending. The other primary options for individuals are imprisonment, home detention, community detention and community work, and these are quintessentially criminal responses to an unlawful act. The ability to imprison offenders had previously only been a feature of a limited number of resource management statutes (see, for example, section 79 of the River Boards Act 1908 and section 62 of the Toxic Substances Act 1979), but with the RMA it was now provided for in every case (along with other non-monetary sentences). The Resource Management Law Reform paper on enforcement and compliance issues had made three relevant points about this. First, it was suggested that a maximum of three months' imprisonment "often finds favour" with policy-makers because it does not give the defendant the right to elect trial by jury pursuant to section 66 of the Summary Proceedings Act 1957 (then in force) (1988a, 46). Second, as noted above, it was suggested that if imprisonment was a sentencing option it would need to be made clear what fault element was required (1988a, 46). Finally, it was said that any "suggestion that imprisonment be available for serious criminal environmental offences is likely to be strongly resisted" (1988a, 46). However, as can be seen, a maximum penalty of two years' imprisonment was ultimately chosen and, as also noted in the section above, the Government chose not to include a fault element for any of the breaches. The resistance to imprisonment, if it ever eventuated, was obviously unsuccessful.

However, as shown in Chapter 4, in practice imprisonment and the other non-monetary penalties are rarely used, meaning that such penalties do little to bolster the criminal character of the RMA's offences. The starting point is that no one was sentenced to imprisonment for RMA offending until 2004, some 13 years after the Act came into force (*R v Conway* [2005] DCR 243). Then, by 30 June 2017 there had only been three further cases where imprisonment was imposed (and none in the 5 years from 1 July 2012). Further, in two of these cases the defendant was already serving a term of imprisonment, which essentially ruled out alternative options, and in the third it was Mr Conway who again offended. While there was something of a renaissance in 2018, with two sentences of imprisonment imposed, these were on the same defendant and were for very serious, repetitive, offending, which suggests it was more likely an outlier than a change in direction (*R v Lau* [2018] NZDC 1133, appeal against sentence dismissed: Lau*v R* [2018] NZCA 151; R v Lau DC Auckland CRI-2006-004-010786, 10 July 2018, appeal against sentence dismissed: R v Lau [2018] NZHC 2935).

In terms of the other non-monetary sentences, as noted in Chapter 4 between 2012/2013 and 2016/2017 only two people were sentenced to home detention, six people were sentenced to community detention and 11 people were sentenced to community work, out of 413 defendants (Ministry of Justice, Letter to author, 23 February 2018). Indeed, no offenders were sentenced to community-based sentences or home detention in the final year under consideration, 2016/2017. It must, of course, be observed that a good proportion of those convicted—likely around a half—will be non-natural persons, and so imprisonment and the other non-monetary sentences were not available, but even taking this into account these sentences are infrequent.

Indeed, the preference by local authorities for non-natural person defendants, as argued in Chapter 6, is part of the problem, given the lack of available alternatives to a fine when sentencing them. It is also unsurprising, given the splitting (and raising) of the maximum monetary penalty in 2009. The implicit message to local authorities from providing a higher maximum monetary penalty for non-natural persons is that the focus should be on companies, as the recovery may be greater than if they proceed against an individual. The maximum penalty was, apparently, split "to reflect ability to pay and likely commercial motives" (Cabinet Paper 2009, 26), but there was no evidence provided to support the proposition that in RMA cases a non-natural person has a greater ability to pay or is more likely to have a commercial motive than a natural person. While this may be the case in some situations, one can equally identify examples of the opposite—for instance, a charitable trust may have far fewer resources and a more benevolent motivation than many sole traders. Even taking (probably) the most prevalent type of prosecution, that of a dairy farm owner for an effluent discharge, whether the farm is operated by a small family-owned company or a husband and wife partnership, both their ability to pay and their motive is likely to be the same. The only place where a higher maximum penalty for non-natural persons may make sense is to enable local authorities to deal with very large corporations, but these are not prevalent in New Zealand, with 97% of enterprises in New Zealand having fewer than 20 employees (Ministry of Business Innovation and Employment 2017). Further, if this was (part of) the rationale then it has likely not worked—for instance, the fine of $288,000 in *Bay of Plenty Regional Council v Mobil Oil New Zealand Ltd* [2016] NZDC 8903 pales in comparison to the US$16.2b the defendant's parent company earned the year of the offending.

Let us step back for a moment. While historically successful prosecutions were resolved by the practice of paying compensation to the person bringing the case, this was abolished hundreds of years ago and replaced with the payment of fines to the Crown (Klerman 2001, 6). Further, in terms of these fines, it has, as discussed in Chapter 5, traditionally been considered unlawful to insure against them, on the basis that this would be contrary to public policy. However, the RMA does not follow these positions. The vast majority of any fine imposed goes to the party that initiated the proceeding, in what appears to be compensation for the "damage" it has suffered (by way of having to pay investigation and prosecution costs), and insurance against these fines is not prohibited, seemingly to ensure that this compensation is in fact paid. Such responses to offending are distinctly non-criminal, and have become the focus of RMA sentencing over the quintessentially criminal responses of imprisonment and other non-monetary penalties. Indeed, the most recent change to the sanctioning regime (in 2009), increasing the maximum monetary penalty and splitting it so that it is twice as high for non-natural persons, has only served to reinforce this. Should a "secondary" penalty be imposed, as shown in Chapter 4 this is typically a "compliance-focussed" enforcement order, which is a civil remedy.

There is, however, one distinctly criminal consequence of a successful prosecution that usually cannot be avoided, and that is the conviction. A conviction is generally considered to be a penalty, quite apart from any other sentence imposed, because it labels the defendant as a "criminal" (Simester et al. 2013, 5). By saying "X is a murderer", and all the connotations that come with this, this is in fact shorthand for saying "X is a *convicted* murderer". In the usual course, this invites the community to make an adverse moral judgment about the defendant's character—as Simester and Von Hirsch note, "[t]he very labelling of a defendant as 'criminal' imports all the resonance and meaning of that term—quite different from publicly calling her a 'tortfeasor' (!)" (2011, 5).

However, the courts have, for the most part, suggested that an RMA conviction says little about the defendant's character. For instance, in the sentencing decision of *North Shore City Council v Aislabie* DC Auckland CRI-2008-044-003188, 14 October 2008, Judge McElrea stated, when considering an application for a discharge without conviction on RMA charges, that:

> [67] There is no evidence for example, that there is any question wider than one relating to criminal convictions - and *this type of regulatory offence is not a crime*. There is no evidence that there is any bar or any discretion exercised in relation to offences of this type. The same comment applies in relation to being a marriage celebrant. Indeed, I would be most surprised if a conviction for cutting down a tree or trees without consent under the RMA would operate as any bar in that way. It *does not cast a slur on the moral integrity of Mr Aislabie*. (emphasis added)

More recently, Judge Thompson in the sentencing decision of *Taranaki Regional Council v Collingwood* [2016] NZDC 16616 was continuing to promote a similar position when considering an application by one of the defendants for a discharge without conviction:

> [20] It is to be borne in mind, I say finally, on that point that convictions for offences under the Resource Management Act do not carry overtones of dishonesty or immorality or violence. They are regulatory offences, and on this occasion I am afraid Mr Collingwood has lapsed and that is that.

It can be observed that, in 2018, Judge Harland was putting forward a contrary position, in the sentencing decision of *Bay of Plenty Regional Council v Whitikau Holdings Ltd* [2018] NZDC 3850. In particular, Her Honour stated that (at [217]) while "arguably environmental offending was historically seen as more 'regulatory offending', or as Moore DCJ said in [*Auckland Regional Council v Gubbs Motors Ltd* DC Auckland CRN-08088500246/08084500006, 20 March 2009] 'statutory breaches' rather than truly 'criminal'", it is (at [218]):

> … worth exploring this issue a little further, and offering the idea that this distinction is no longer as valid as it once may have been. This is because there is more understanding these days of the impact environmental offending can have on human activity, health and wellbeing, particularly if one takes a longer term and intergenerational view. As well, classifying these offences in this way has the effect of minimising them, and ignoring the maximum penalties provided for such offending by Parliament, and the fact that these are not civil or pecuniary penalties.

However, in this same judgment Judge Harland felt compelled to allow the fines to be insured against, which as noted above, is a distinctly non-criminal response, and Her Honour accepted that the impact of a conviction will be variable (at [230]).

Since that case the regulatory nature of RMA offending has been emphasised by the High Court, with Whata J starting in *R v Lau* [2018] NZHC 2935 that:

> [34] …The present offending is not like violent offending as Mr Symon submits. The requirement for denunciation and deterrence is not the same, as the maximum sentences for violent offending clearly show. It is necessary to keep this firmly in mind when dealing with regulatory breaches of the present kind.

Or as Brewer J put it, when considering an appeal against a refusal to grant a discharge without conviction in *Withington v Bay of Plenty Regional Council* [2018] NZHC 1237:

> [26] First, I shine the light of reality on the situation. Mr Withington pleaded guilty to a charge under the Resource Management Act. He has not lied, cheated, defrauded, been violent or depraved.

The message that is therefore sent by the courts when sentencing offenders, by and large, is that the offending is "only regulatory", and convictions carry no adverse moral judgment. Convictions are, rather, simply what enables the Court to impose a fine, as if a defendant is discharged without conviction then no other primary sentence can be imposed (Sentencing Act 2002, section 106(3)). As such, it can be seen from this section that what is missing in the RMA are the "two central functions" of the criminal law, the condemnation from the conviction and punishment (Tadros 2010, 164–165).

Sui Generis Institutional Arrangements for Prosecuting

By taking an expansive approach to the offences and focussing penalties on compensating local authorities, prosecuting *could* be made easier and more worthwhile, but what was also required to ensure this happened was the creation of a receptive institutional environment. The Government endeavoured to achieve this in various ways, the first of which was by placing no restrictions on how local authorities were to undertake RMA prosecutions.

There was limited mention of who would be primarily responsible for prosecuting RMA offences, and how this would be done, during the Resource Management Law Reform process. For instance, the working paper on enforcement and compliance issues simply stated that:

> Some issues arising out of legislation include:
>
> Who can prosecute? Under S.13 of the Summary Proceedings Act generally can "lay an information". In other Acts, there are specified exemptions where only particular people can lay charges. (Generally some wrong needs to done to them but the degree of effect varies).

> The Clean Air Act is enforced by the local authority but some prosecutions need an officer appointed or the leave of the Director-General Health. Under the Water and Soil Conservation Act anyone can prosecute, but only the Board can request an injunction pursuant to S.34(6). (Resource Management Law Reform Core Group 1988a, 38).

This was, probably, a reflection on the fact that the responsibilities for implementing (and so enforcing) the new legislation were still being resolved when this working paper was issued. However, even when the Act was passed it was never explicitly stated that it was the role of local authorities to undertake prosecutions—as mentioned in Chapter 5, the only indication that this is the case is the somewhat ambiguous requirement in section 84(1) of the Act that "[w]hile a policy statement or a plan is operative, the regional council or territorial authority concerned, and every consent authority, shall observe and, to the extent of its authority, enforce the observance of the policy statement or plan". Section 342 (fines to be paid to local authority instituting prosecution) arguably provides the clearest indication that local authorities will be undertaking prosecutions, but this could equally be read as simply recording what is to happen if they do prosecute (rather than a responsibility to do so).

This is important because local authority prosecutions, in general, have never been adequately assimilated into the criminal law infrastructure in New Zealand. For instance, when the Law Commission considered criminal prosecutions (generally) in the late 1990s, council matters barely featured and were considered in a confused manner. In the Law Commission's Preliminary Paper it referred to "local bodies" under the heading "Other Government Prosecuting Agencies", and after the comment that "Apart from the police and the Serious Fraud Office, a range of other public bodies prosecute offences", but it did not consider their prosecution practices (Law Commission 1998, 29). In its final report on Criminal Prosecution, however, the Law Commission was classifying local authorities (what it termed "local and quasi-public bodies") as private prosecutors, although it observed (in a footnote) that they may be "characterised in another sense as public not private, despite not using the police or Crown solicitors, as their overall function is public" (Law Commission 2000, 92).

This uncertainty over whether local authority prosecutions are public or private carried over to the Criminal Procedure Act 2011. The starting point is that local authority prosecutions are explicitly excluded, in section 5 of that Act, from the definition of a "private prosecution". This is probably so that the checks on private prosecutors, such as restrictions on the filing of charging documents, do not apply. However, local authority prosecutions are also not "public prosecutions"—section 5 of the Act defines these as "a proceeding in respect of an offence that is commenced by or on behalf of the Crown, and includes a proceeding in respect of an offence that is commenced by or on behalf of a Crown entity within the meaning of section 7 of the Crown Entities Act 2004". The main consequence of this is that local authority prosecutions do not come within the "general oversight" of the Solicitor-General (Criminal Procedure Act 2011, section 185), and it is not mandatory for councils to apply the *Solicitor-General's Prosecution Guidelines* (Crown Law Office 2013).

The fact that this is not simply a mistake, given that the classification of local authority prosecutions was obviously considered when the definition of "private prosecution" was determined, suggests instead that such prosecutions are considered, from a criminal infrastructure standpoint, unimportant. Put another way, private prosecutions are generally considered "a useful constitutional safeguard against capricious, corrupt or biased failure or refusal of those authorities to prosecute offenders against the criminal law" (*Gouriet v Union of Post Office Workers* [1978] AC 435 at 498), but since they can also be used problematically, for instance as a form of harassment, there are restrictions on them. On the other hand, public prosecutions reflect that it is the community that is generally responsible for punishing wrongdoing, but as a matter of fairness to defendants (and putative defendants) there must be some central oversight to ensure that such prosecutions are being conducted consistently. By not allocating local authority prosecutions to either category, the implication is that RMA prosecutions (and other local authority prosecutions) do not have a specific role in the criminal law landscape.

In a similar vein, because local authorities are not government departments, the Cabinet Directions for the Conduct of Crown Legal Business 2016 do not apply to them (Cabinet Office Circular 2016, Annex 1).

This means that, in addition to the Police not being involved, a local authority does not need to engage a firm headed by a Crown Solicitor to run its prosecutions, and can instead choose to be represented by in-house counsel, a private law firm or a barrister. By the same token, local authorities will be responsible for paying for this representation, and there is no restriction on the charging of market rates even if a Crown Solicitor's office is engaged. There may well be practical benefits in local authorities having this choice, in that regular counsel may build up specific expertise in RMA prosecutions, but the impression, once again, is that local authority prosecutions do not fit into the usual criminal mould.

There are also practical problems with the combination of a lack of identification of local authority prosecutions as private or public and that local authorities can choose who they want to represent them in court. For instance, because a local authority prosecution is not a "private prosecution", then pursuant to section 5 of the Criminal Procedure Act 2011 a prosecution brought by a local authority will become a "Crown prosecution", and so be taken over by the local Crown Solicitor's firm, in two situations. One is if the defendant elects to be tried by a jury (Crown Prosecution Regulations 2013, regulation 4(1)(c)), and the other is if the Solicitor-General directs that, having regard to the particular features of the proceeding, it is appropriate that it be conducted as a Crown prosecution (regulation 4(1)(e)). While the latter is rare, the former occurs from time to time (and the frequency seems to be increasing), and it can potentially lead to confusion, or at least double-handling, if a local authority instructed a private law firm to act in first instance, only for the matter to be handed over to a Crown Solicitor's firm after the election of a jury trial some months later. The *Solicitor-General's Prosecution Guidelines* will then also apply and "all decisions in relation to disclosure, the charges filed, the evidence to be adduced, the conduct of the prosecution and the nature and scope of any continuing investigation (where it is probable that will result in evidence or information relevant to the trial) are matters solely for the Crown prosecutor to decide", although "Crown prosecutors are expected to consult closely with and take into account the views of the investigator or officer in charge of the case and to explain the basis of any significant decision" (Crown Law Office 2013,

28.29–28.10). Further, if the *defendant* has instructed the local Crown Solicitor's firm to act for it, on the basis that the local authority was prosecuting the matter itself or instructing a private law firm or barrister, and then elects trial by jury, this creates issues for who will act for the Crown. This occurred in *Greymouth Petroleum Limited v The Solicitor-General of New Zealand* HC Wellington CIV-2009-485-1425, 25 February 2010, with the defendant unsuccessfully judicially reviewing decisions of the Solicitor-General to file an indictment against it, and to refuse to grant dispensation to the New Plymouth Crown Solicitor to continue to act as its counsel at trial.

Turning to the proceedings themselves, it is, of course, the case that RMA prosecutions are heard in the District Court, like most other criminal offending in New Zealand. However, section 309(3)(b) of the RMA provides that, except where otherwise directed by the Chief District Court Judge, the prosecution will be heard by "a District Court Judge who is also an Environment Judge". This was not a feature in the previous legislation, although it seems that, where possible, relevant cases were being referred to Planning Judges (as they were called then) (Resource Management Law Reform Core Group 1988b, 33). A review of the Town and Country Planning Act 1977 had recommended that criminal cases should be (solely) adjudicated by such a specialist Judge (Hearn 1987, 106), and this was supported in the working paper on enforcement and compliance issues (Resource Management Law Reform Core Group 1988a, 58). On the other hand, the recommendation in the Resource Management Disputes working paper, which the Ministry for the Environment did not appear to disagree with (although issues with it were raised), was that there should be concurrent jurisdiction between the District Court and the Planning Tribunal (judge sitting alone), with a power vested in the Chief Planning Judge to transfer a case from one to the other (Resource Management Law Reform Core Group 1988b, 34). The outcome was probably the same, however, that Environment Judges, as opposed to "generally warranted" District Court Judges, would typically hear the prosecutions, and this was what was enacted.

This is understandable given Environment Judges have specialist knowledge of environmental law and the RMA, but such Judges are more commonly associated with their civil work. For instance, Environment

Judges are, first and foremost, part of the Environment Court, with prosecutions falling into their "other" duties. This can, to an extent, be seen in the fact that for the year ended 30 June 2017, 453 cases were disposed of in the Environment Court (Registrar of the Environment Court 2017, 9–10), while only 71 prosecutions were commenced (and the majority of these would have involved only administrative appearances or sentencing hearings, given how few cases go to trial) (Ministry for the Environment 2018b). More tellingly, perhaps, is that the Environment Judges' Annual Review for 2017 devoted only two paragraphs in the 32 page report to RMA prosecutions, suggesting these are considered by the Judges to be a lesser part of their role (Environment Court of New Zealand 2017, 10).

Further, when Environment Judges deal with RMA prosecutions, these are often not treated like mainstream criminal cases. For instance, RMA prosecutions, other than when they are first called before a registrar, are not dealt with in "general" lists, like most District Court criminal cases, but are given their own hearings and management. In hearings, Environment Judges at times take a more active role, for instance by trying to resolve (what they consider to be) questionable prosecutions (for instance, *Auckland Council v Liu* [2016] NZDC 14243, as referred to in Chapter 6), or by suggesting uses for a fine in proceedings brought against a territorial authority (for instance, *Otago Regional Council v Queenstown Lakes District Council* [2017] NZDC 28767), or by asking counsel to file additional submissions when they are do not consider the starting point fine sought to be high enough (for instance, *New Plymouth District Council v Cowley* [2019] NZDC 21662). Indeed, in *Bay of Plenty Regional Council v Mobil Oil New Zealand Ltd* [2016] NZDC 8903 the sentencing Judge went so far as to say that, because the fine imposed in that case would not be enough to meet the prosecutor's costs, "there may need to be some review of the legislation in due course to take into account that these costs are otherwise borne by the rate payers through no cause of their own" (at [59]–[60]).

It is worth looking at one example in more detail, and this resulted from the aftermath of the decision that it is possible to insure against RMA fines, *Bay of Plenty Regional Council v Whitikau Holdings Ltd* [2018] NZDC 3850. After this case, a practice developed whereby defendants were being asked by the courts if they carried insurance, and if so the matter was adjourned for the defendant to provide: confirmation as to the scope of the indemnity provided to the defendant under the policy; the amount of cover of any excess, and how the policy interprets that it is to apply; what the policy excludes that might be relevant to the particular offending upon which the sentence is to be based; and any other relevant conditions of cover that might be relevant to the assessment of culpability for the offending. This is unusual for a typical criminal sentencing, where the relevant question is simply whether the defendant can pay a fine of the suggested level, and in *H & S Chisholm Farms Ltd v Waikato Regional Council* [2018] NZHC 1885 the High Court held that such an enquiry about insurance must take place within the specific terms of the Sentencing Act 2002. That is, the sentencing court must first be uncertain about the defendant's ability to pay any fine that may be ordered (section 41(2)), and only if so can it request a declaration as to financial capacity (section 42), which can include information about insurance held against fines imposed under the RMA.

This has greatly restricted the ability of the District Court to ask for insurance information, as it will often be the case that there is no issue as to the defendant's ability to pay a fine. For instance, even if the insurance company has not finally decided whether to accept cover then, so long as the defendant has sufficient resources of its own it can confirm to the Court it can pay a fine of the level suggested and there should be no need to consider the matter further. The key point, though, is that this shows that Environment Judges tried to come up with a non-criminal way (enquiring into who exactly would be paying the fine) of resolving a non-criminal outcome of offending (insurable fines), but the criminal law (in the form of the Sentencing Act 2002) is not amenable to such use. Put another way, the focus of the provisions of the Sentencing Act is on whether the fine will be paid, not by whom, as it was probably never contemplated that insurance would be an issue it needed to address.

Ultimately, RMA prosecutions are being heard by judges that spend the majority of their time dealing with the civil work of the Environment Court, and the cases themselves are isolated from the general criminal law work undertaken in the District Court and seem to involve more active judicial involvement than one would typically expect. Indeed,

RMA prosecutions are treated quite differently from even other regulatory offending, which is dealt with by generally warranted District Court Judges in "criminal" lists. This point is primarily one of perception—that hearings for RMA prosecutions do not look "criminal"—but parallels can be drawn with the decision of the New South Wales Land and Environment Court to wear robes, after initially deciding not to, as explained by Pearlman:

> When the Court was established, a rule was promulgated that the judges would not robe. But as cases became more complex, and, more importantly, as the penalties for environmental offences became more severe, the judges felt that there was a need to reinforce the status of the Court as a superior court of record, especially in relation to environmental prosecutions. Accordingly, in 1998, the judges resolved to robe when sitting in cases in those classes of the Court's jurisdiction… (2000, 4)

Walters and Westerhuis note that while "this may seem a superficial shift, this was a different strategy to emphasise the seriousness of environmental crime, following the more overt emphasis on prosecution of environmental crime a decade before" (2013, 285). Further, it may be that this also explains why, as set out in Chapter 6, the courts seem reluctant to stop local authorities using the offences in a non-criminal way, except in the most extreme cases.

Conclusion

The implication from the foregoing is that the reason local authorities consider it acceptable to act in an inappropriate way is because the offences, together with the penalties and institutional arrangements for prosecuting, contain few traces of "criminality". The offences go far beyond sanctioning only serious wrongdoing (by reference to harm and culpability), they rarely result in quintessentially criminal penalties and convictions convey no adverse moral judgment. Further, the offences are dealt with, and penalties imposed, in an environment where prosecutions have no special oversight from the courts or Solicitor-General, and cases are heard separate to other criminal matters and managed accordingly.

As such, it is no surprise that, as concluded in Chapter 6, the approach taken by councils is not only distinctly "non-criminal", but is in fact "civil", as the RMA's offences, penalties and institutional arrangements are, in substance, more akin to those found with civil causes of action. As Yeung has noted, civil law remedies:

> ...do not paradigmatically involve the punishment of the wrongdoer, nor are they associated with the degree of moral stigma typically accompanying a finding of criminal liability. Because the consequences of civil liability for the wrongdoer are considerably less serious than those associated with criminal wrongdoing, proof of culpability need not be insisted upon in defining the relevant standards of liability, and the procedural rights of the suspected wrongdoer are considerably weaker than those available to the criminal defendant. (2004, 250)

Such features of the civil law—and more—are present here. The offences allow for a wide range of parties to potentially jointly cause a breach, intention is irrelevant and the defendant has to disclose and prove his or her defence. Further, a successful case typically results in a morally neutral monetary penalty, which can be insured against, being returned to the party bringing the proceeding. And finally, this all takes place in a system where the initiator can instruct (and pays for) its counsel of choice, and the matter is heard by Judges who spend most of their time doing non-criminal work.

This is an appropriate point at which to conclude the assessment of the RMA's offences. In the next chapter, the last in this book, we look at what lessons can be learned from the case study.

References

Cabinet Office Circular. 2016. Cabinet Directions for the Conduct of Crown Legal Business 2016.

Cabinet Paper. 2009. Reform of the Resource Management Act 1991: Phase One Proposals.

Campbell, Janette. 2015. Statutory Remedies: The Enforcement Provisions of the Resource Management Act 1991. In *Environmental and Resource Management Law*, ed. Derek Nolan. Wellington: LexisNexis.

Crown Law Office. 2013. Solicitor-General's Prosecution Guidelines.

Environment Court of New Zealand. 2017. Annual Review Calendar Year 2017.

Farrier, David. 1992. In Search of Real Criminal Law. In *Environmental Protection and Legal Change*, ed. Tim Bonyhady, 79–124. Sydney: The Federation Press.

Grinlinton, David. 1992. Enforcement Mechanisms Under the RMA. *Planning Quarterly*: 15–19.

Hearn, Antony. 1987. Review of the Town & Country Planning Act 1977: Department of Trade and Industry.

Horder, Jeremy. 2014. Bureaucratic 'Criminal' Law: Too Much of a Bad Thing? In *Criminalization: The Political Morality of the Criminal Law*, ed. R.A. Duff, L. Farmer, S. Marshall, M. Renzo, and V. Tadros, 101–131. Oxford: Oxford University Press.

Klerman, Daniel. 2001. Settlement and the Decline of Private Prosecution in Thirteenth-Century England. *Law and History Review* 19 (1): 1–65. https://doi.org/10.2307/744211.

Law Commission. 1998. Criminal Prosecution: NZLC PP28.

Law Commission. 2000. Criminal Prosecution: NZLC R66.

Ministry for the Environment. 2018a. Best Practice Guidelines for Compliance, Monitoring and Enforcement Under the Resource Management Act 1991.

Ministry for the Environment. 2018b. National Monitoring System for 2016/17.

Ministry of Business Innovation and Employment. 2017. Small Businesses in New Zealand.

Palmer, Geoffrey. 2013. The Resource Management Act—How We Got It and What Changes Are Being Made to It. Paper presented at the Resource Management Law Association, New Plymouth, 27 September.

Palmer, Geoffrey. 2016. Ruminations on the Problems with the Resource Management Act—Part 2. *New Zealand Law Journal* 46–49: 80.

Pearlman, Mahla, Chief Judge of the Land and Environment Court of New South Wales. 2000. The Land and Environment Court of New South Wales. Paper presented at the Royal Australian Planning Institute Congress, Sydney, 5 October

Registrar of the Environment Court. 2017. Report of the Registrar of the Environment Court for the 12 Months Ended 30 June 2017.

Resource Management Law Reform Core Group. 1988a. Enforcement and Compliance Issues in Resource Management: Ministry for the Environment. Working Paper 30.

Resource Management Law Reform Core Group. 1988b. Resource Management Disputes: Part A: The Role of Courts and Tribunals Part B: Mediation: Ministry for the Environment. Working Paper 22.

Simester, A.P., J.R. Spencer, G.R. Sullivan, and G.J. Virgo. 2013. *Simester and Sullivan's Criminal Law: Theory and Doctrine*, 5th ed. Oxford: Hart Publishing.

Simester, A.P., and Andreas Von Hirsch. 2011. *Crimes, Harms, and Wrongs: On the Principles of Criminalisation.* Oxford: Hart Publishing.

Tadros, Victor. 2010. Criminalization and Regulation. In *The Boundaries of the Criminal Law*, ed. R.A. Duff, L. Farmer, S. Marshall, M. Renzo, and V. Tadros, 163–190. New York: Oxford University Press.

Waikato Regional Council. 2016. Basic Investigative Skills for Local Government.

Walters, Reece, and Diane Westerhuis. 2013. Green Crime and the Role of Environmental Courts. *Crime, Law and Social Change* 59 (3): 279–290. https://doi.org/10.1007/s10611-013-9415-4.

Yeung, Karen. 2004. *Securing Compliance: A Principled Approach.* Oxford: Hart Publishing.

Part IV

Concluding

8

Lessons Learned

We have seen, in the previous four chapters, that the RMA's offences are neither effective nor being used appropriately. In terms of effectiveness, despite the primary purpose of the offences being typically said to be deterrence, it is unlikely that would-be offenders are being deterred. This is because the penalties imposed pursuant to the offences are of no more than moderate severity and there is a low certainty that such a penalty will be imposed in response to an RMA breach. Together, this gives the offences a minimal expected cost, and so it would only take a correspondingly low expected benefit from breaching the RMA for a rational regulatee to offend. In terms of appropriateness, the offences should be used in a manner that complies with well-established criminal law values, such as equality of treatment and fairness. However, we have seen that, at times, local authorities act contrary to these values, in the way they decide what types of charges to bring, in the way they "settle" prosecutions in court, and in the way they resolve cases out of court.

It is now time to turn our attention to what lessons can be learned from the case study, of which there are three. First, the case study has shown the importance of aligning the substance of any offences, penalties and institutional arrangements for prosecuting with their criminal form.

M. Wright, *Responding to Environmental Crimes*, Palgrave Studies in Green Criminology, https://doi.org/10.1007/978-3-030-89250-0_8

Second, it has shown that collecting good data on the use of offences requires an understanding as to how the criminal law works. Third, and perhaps most importantly, the case study raises serious questions about the ability of the criminal law *in general* to deter. Each of these lessons is discussed in this chapter, before the book ends with some final comments on the implications for dealing with environmental wrongdoing and green criminology.

Lesson 1—Align the Form and Substance of Offences, Penalties and Institutional Arrangements for Prosecuting

Policy-makers worldwide have a tendency to stray from criminal law norms when developing offences, penalties and institutional arrangements for prosecuting, in order to make things easier for those enforcing environmental regulation statutes (see generally O'Hear 2004). However, it is suggested that the fundamental reason the RMA's offences are both ineffective and being used inappropriately is that this tendency has been taken to the extreme in New Zealand, with the RMA's offences, penalties and institutional arrangements for prosecuting being, in substance, more akin to those found in civil matters. The first lesson from the case study, therefore, is that it is important that the form and substance of offences, penalties and institutional arrangements for prosecuting are aligned.

Let us start where we left off at the end of Chapter 7, with the conclusion (referred to above) that the RMA's offences, penalties and institutional arrangements for prosecuting are, in substance, more akin to those found in civil matters. In particular, reference was made to the wide range of parties that could jointly cause a breach, the irrelevance of intention and the requirement on the defendant to disclose and prove his or her defence. It was also observed that a successful case typically results in a morally neutral monetary penalty, which can be insured against, being returned to the party bringing the proceeding. Finally, it was shown that this all takes place in a system where the party bringing proceedings can instruct (and pays for) its counsel of choice, and the matter is heard by judges who spend most of their time doing non-criminal work.

It is this civil substance that has encouraged local authorities to adopt (what was termed in this book) the "compensatory gloss" approach to the offences, whereby what is most important is that any damage caused by the offender is remedied (or the issue that led to the non-compliance is fixed) and the council recovers as much of its costs incurred as it can, but such an approach is causing inappropriate behaviour (see Chapter 6). For instance, it is leading local authorities to focus on the most visible breaches and those that are more cost-effective to prosecute, which ends up treating regulatees unequally. It also encourages councils to plea bargain prosecutions in order to get guilty pleas (and agreed summaries of facts) by non-natural person defendants, yet at times this can only be achieved by unfair or unprincipled actions. Finally, it results in local authorities "settling" prosecutable breaches directly, and privately, with some transgressors, which raises questions of suitability (given the seriousness of the offending) and, in some circumstances, may be unlawful.

The compensatory gloss approach is also contributing to the offences being ineffective, with it limiting both the severity of the penalties and the certainty that they will be imposed (see Chapter 5). In terms of severity, it means that local authorities prefer the compensatory sentence of a fine to sterner alternatives such as community work, imprisonment or an order that the offender's resource consent be reviewed (which can lead to cancellation of the consent). In terms of certainty, the compensatory gloss approach means that, on the whole, local authorities are reluctant to prosecute due to the cost versus recovery difference involved, especially when compared to other enforcement actions.

On the other hand, the formal criminality of the offences, penalties and institutional arrangements for prosecuting has meant that other parties are unable to increase the effectiveness of the offences. For instance, as discussed in Chapter 5, the courts have rarely imposed more restrictive penalties than a fine (such as "punitive" enforcement orders), other than the few times these have been sought by the local authorities, because criminal law orthodoxy is such that the courts do not typically impose a harsher sentence than that requested by the prosecutor. In a similar vein, because the offences are formally criminal the public has been unable to increase the certainty that a penalty will be imposed. The criminal justice system generally restricts the public's role to making complaints, and as seen in Chapter 4 there is no guarantee in the RMA that these will be followed up by local authorities, with the ability to bring a private prosecution being more theoretical than real.

The formal criminality has also meant that there are no checks on the actions of the local authorities when it comes to using the offences appropriately. As discussed in Chapter 6, because the criminal justice system relies so heavily on prosecutorial discretion the courts have been reluctant to intervene in plea bargains, so long as "someone is pleading guilty to something", and they have frequently allowed local authorities to use diversionary arrangements that rely on charges existing in the background, as a threat. Opportunities for the public to hold local authorities to account are also rare, with only minimal scope provided to challenge a decision of an enforcement agency not to prosecute.

What we are left with, then, is the worst of both worlds. Local authorities have been convinced by the civil substance of the offences, penalties and institutional arrangements for prosecuting that they should not actually be pursuing deterrence and do not need to use the offences appropriately. And the criminal form is stopping others from making up for this lack of deterrence or acting as a check on the actions of local authorities. If policy-makers wish to employ the criminal law to sanction non-compliance then they need to have the courage of their convictions, and ensure that the offences, penalties and institutional arrangements for prosecuting comply with criminal law norms. Whether they *should* be employing the criminal law so frequently is another matter, however, a point that is returned to later in this chapter.

Lesson 2—Understand the Criminal Law in Order to Collect Good Data

So much for the problems caused by the disconnect between the criminal form of the offences, penalties and institutional arrangements for prosecuting and their civil substance, but why have these issues not come to light previously? It is suggested that the reason for this is that the Ministry for the Environment has failed to collect good data on the use of the offences, and so it has been difficult to undertake meaningful analysis. And while a failure to collect good data is a concern that is raised time and again by green criminologists (see, for instance, Nurse 2015, 179; White 2010, 379), what the case study has shown is that obtaining good data on the use of offences requires an understanding of how the criminal law works, which is something that has been lacking here.

Historically, it was difficult to determine precisely how local authorities were using the RMA's offences. While the Ministry for the Environment conducted yearly (and then biennially) surveys of local authorities (from 1995 to 2014), and published a series of prosecution studies (in 2002, 2006, 2009 and 2013), these each had limitations. For instance, while the local authority surveys always contained a section on compliance, monitoring and enforcement, the questions that were asked changed so often and were so vague that the information was of little use. And while the prosecution studies were more comprehensive, they focussed on the end results.

With the creation of the National Monitoring System (which started with the 2014/2015 financial year), the data on the use of the RMA's offences by local authorities became more consistent, and more useful. For instance, during the period that fell within the case study (2014/2015 to 2016/2017) the Ministry counted, each year, the number of recommendations to prosecute, the number of decisions to prosecute and the number of prosecutions initiated. Further, they counted each of these by section of the Act and by local authority, which allowed consideration of the various types of offending and regional differences.

However, this improvement only goes so far. The difference between recommendations to prosecute and decisions to prosecute is hard to discern at times, as it depends on how decisions are made in a council, and these could certainly have been counted together (the figures in these columns are, almost always, exactly the same). More importantly, while the number of times local authorities used the offences each year is worth counting (what the Ministry has called "prosecutions"), the description ("The number of charges laid by your local authority") is confusing, as one prosecution may have many charges. It is also open to interpretation: Must the defendants be charged jointly to count as a prosecution? What if further defendants were added later? What if the charges subsequently get severed from one another? There is also no record of how many defendants were charged and how many charges they faced, which is important information when it comes to *how* the prosecutions were resolved (see Chapter 6). In a similar vein, there were also no questions asked of local authorities as to the results of these prosecutions.

It must be observed that, since the case study ended, the Ministry for the Environment has begun counting the number of convictions secured, the number of charges withdrawn and the number of charges acquitted (Ministry for the Environment 2019). This is a further step in the right direction, although the questions could be worded better (for instance, it is *defendants* that are acquitted of charges, not charges acquitted in and of themselves) and it is still not enough. Matters such as the penalty imposed in addition to the conviction are not recorded and, indeed, it is not clear what local authorities are to do if the defendant is discharged without conviction under section 106 of the Sentencing Act 2002. More broadly, there is no link between the prosecutions initiated and the convictions/withdrawals/acquittals (which may have been in a different year). Given the prosecution studies, which used to fill this gap somewhat, have apparently been discontinued (Brown 2017, 53), the National Monitoring System does not actually take us much further than the methods of recording previously used, with a lot of work required to undertake any proper analysis on the use of the offences.

The short point is that, despite the RMA being in existence for 30 years now, the Ministry for the Environment is still not quite sure what it should be counting in terms of the use of the offences, and the implication from the above is that this is because it does not understand how the criminal law works. Indeed, this is as good as revealed by the "Message from the Minister" in the Ministry for the Environment's *Best Practice Guidelines for Compliance, Monitoring and Enforcement under the Resource Management Act 1991*, where he notes that "[e]nforcement of the rule of law will always be essential to encourage broader compliance. *This is true in criminal*, transport, taxation, *or environmental law*"

(2018a, 7 (emphasis added)). As far as the Ministry is concerned, the offences in the RMA, despite being governed by (among others) the Criminal Procedure Act 2011, the Criminal Disclosure Act 2008, the Sentencing Act 2002 and the Costs in Criminal Cases Act 1967, are not, apparently, a matter of criminal law, but rather environmental law.

Lesson 3—The Criminal Law Struggles to Deter

Making sure that the substance of offences, penalties and institutional arrangements for prosecuting aligns with their criminal form, and having an understanding of how the criminal law works so that good data is collected on the use of the offences, would have gone some way to preventing the problems and/or identifying them when they occurred. But there is a third lesson from the case study, which is perhaps the most important. This lesson is that, while it is generally assumed that criminal penalties are a good deterrent (Preston 2007, 95), in fact offences will most often struggle to deter. And while this may not appear to be a particularly novel finding (see, for instance, Faure and Svatikova 2012), what is of note here is just how considerable the struggle is and typically will be.

It will be recalled, from Chapter 5, that the reason the fine levels in the RMA are only modest is, essentially, because judges are being constrained by proportionality requirements and they are having to give offenders a number of "discounts". For instance, although the Sentencing Act 2002 sets out all rationales of punishment in section 7 as available options, the principles of sentencing in section 8 suggest that, when actually imposing the penalty, proportionality (to seriousness) should be the overriding concern (Roberts 2003, 259). Further, the standard approach to criminal sentencing, which has been adopted in the RMA (*Calford Holdings Ltd v Waikato Regional Council* (2009) 15 ELRNZ 212 (HC)), involves a judge setting a starting point for the offending—"the sentence considered appropriate for the particular offending (the combination of features) for an adult offender after a defended trial" (*R v Mako* [2000] 2 NZLR 170 (CA) at [34])—and then making adjustments for personal aggravating or mitigating factors, the latter including matters such as remorse, previous good character and, most importantly, a reduction for any guilty plea (Sentencing Act 2002, section 9(2)).

Importantly for present purposes, however, such restrictions on penalties are not unique to New Zealand, but are instead a common feature of the criminal law. For instance, in the English context, despite the sentencing legislation in that jurisdiction also containing a number of purposes of sentencing, the Sentencing Council's "Overarching Principles" focus on proportionality (Ashworth 2010, 78). The same applies to discounts, and in particular the deduction for a guilty plea, which is increasingly being given statutory prominence (Welsh et al. 2021, 358–361). Put shortly, therefore, tying sentences to the seriousness of the offending and reducing penalties to take into account the defendant's individual circumstances is an entirely orthodox approach, and so any sanctions for environmental non-compliance that rely on the criminal law will typically also be constrained by these features.

A common response to offences being insufficiently severe is to increase maximum penalties, but this was tried with the RMA and did not work. In particular, the increase did not flow down to sentence starting points in a dollar-for-dollar manner, and so penalties did not rise significantly. It would, of course, be possible to enact a very large increase in the maximum penalty (to achieve the required proportionate increase), but this would raise legitimacy concerns when these maximum penalties were compared to those for offences in other statutes. Further, care would need to be taken to ensure that any increase in severity was not balanced out by a corresponding decrease in certainty, as enforcers become complacent that a few exemplary sentences mean they do not have to prosecute as often. This seems to have happened in New Zealand, with the number of prosecutions brought decreasing after the maximum penalty was raised (see Chapter 4).

Turning then to squarely address certainty, which is generally considered the more important of the two deterrence factors, the case study has shown that local authorities are reluctant to prosecute. We saw in Chapter 4 that in 2016/2017, the last year considered in the case study, that out of the 3,344 formal enforcement actions taken a mere 71 were prosecutions (Ministry for the Environment 2018b), and indeed formal enforcement action itself was only taken in response to a fraction of the detected breaches (an unknown number, but likely to be in the tens of thousands). This year was by no means an outlier either, with only 57 prosecutions and 81 prosecutions commenced in 2015/2016 and 2014/2015 (respectively) (Ministry for the Environment 2016, 2017).

The importance of this is that, despite the way that environmental non-compliance is enforced in New Zealand (local authorities) being different from that found overseas (often, environmental protection agencies (see Chapter 2)), what is common throughout the various jurisdictions is the use of the criminal law as the ultimate sanction, and the case study has shown that even extremely favourable offences, penalties and institutional arrangements could not motivate New Zealand's enforcers to bring many prosecutions. As discussed in Chapter 7, despite every offence being strict liability, the liability net being cast wide, defences being limited and needing to be advised to the local authority in a very short timeframe, fines being returned to the prosecuting council, local authorities being able choose who they want to represent them and judges often taking an active role in the management of cases, this is still not enough. All of which suggests that the problem is actually with the employment of the criminal law itself.

It must also be remembered that deterrence relies on the awareness of regulatees as to severity and certainty (the "expected" cost of offending), and yet the RMA experience has been that environmental prosecutions do not lend themselves to lasting publicity. It is of course the case that some successful prosecutions result in a press release and/or media attention, but there is rarely any ongoing record that would give the regulated community any idea of the prospect that they will be prosecuted and, if so, what penalty they might get. The most obvious place to provide this would be on the relevant local authority's website, but there has been a reluctance to place details of successful prosecutions there (see Chapter 4). On the other hand, some local authorities are prepared to advertise on their websites the names of those issued with an infringement notice, the low-level punitive measure (see, for example, Environment Canterbury 2021). There is something perverse in this—that the low level offenders are "named and shamed" on a continuing basis while the more serious offenders are forgotten—and the best explanation is that the restrictions that may be placed on the reporting of criminal offending (although rarely are in RMA matters) have made the local authorities unwilling to provide this information in an enduring manner.

Overall, it therefore appears to be the invocation of the criminal law itself that is standing in the way of deterrence. The courts will generally be constrained by criminal law orthodoxy from increasing severity to the levels required, and this cannot be helped by simply raising the maximum penalty. In terms of certainty, enforcers are reluctant to prosecute frequently, even when the offences, penalties and institutional arrangements are heavily in their favour. Indeed, the reluctance of councils to publicise successful prosecutions in any ongoing way suggests that the criminal law, rather than being a powerful weapon that those tasked with enforcement are fortunate to have at their disposal, is instead something they would prefer to not have to deal with.

Implications and Concluding Comments

What are the implications from the lessons above? As a starting point, it is suggested that if deterrence is the sanctioning goal then more consideration needs to be given to non-criminal alternatives, which are not necessarily shackled by things such as discounts for guilty pleas and which may prove more agreeable to those enforcing the law. It is, however, important to be clear that this is quite different from saying that the criminal law should not be used at all for environmental wrongdoing; indeed, the analysis conducted in Chapter 3 suggests that sufficiently harmful and culpable conduct could, and indeed should, be dealt with by the criminal law. Rather, it would likely mean a split approach—a broad array of non-criminal breaches that are sanctioned to deter and a narrow set of criminal offences where the purpose is to condemn.

There would be nothing inappropriate about a split approach. As Ashworth and Zedner have noted, "[s]ome wrongs are so egregious that they ought to be subject to the censure of the criminal law and state punishment. Other, lesser wrongs may properly, and without undermining significant social values, be dealt with in some kind of lesser system of administrative offenses or civil wrongs" (2012, 552). And while Ashworth and Zedner go on to note that dividing wrongs up between the two may be difficult, this should not stop us from attempting the task—taking the easy approach and simply criminalising everything is part of the reason that the offences in the RMA are so problematic.

If the criminal law is to be employed only for sufficiently harmful and culpable conduct, then it naturally follows that the offences that deal with such behaviour should be drafted in a manner that reflects this. In a similar vein, it is important, when determining the available penalties, to avoid perverse sanctioning incentives, such as we have seen with the RMA. Some consideration should also be given to institutional arrangements for prosecuting, which are no less important but are frequently considered secondary to offences and penalties. For instance, are specialist judges needed? The reason commonly cited for requiring this in environmental cases is that specialist judges are better placed to understand the nature of the harms caused (see generally White 2013b). However, if only serious environmental offending is dealt with by the criminal law then we have perhaps reached the stage where this is no longer as necessary. Indeed, it can be observed that when RMA prosecutions come before New Zealand's senior—generalist—courts, on the basis that the defendant is arguing that the sentence imposed was manifestly excessive, these appeals are rarely allowed, and indeed it is sometimes suggested by these courts that harsher penalties would have been appropriate (see, for example, Lau *v* *R* [2018] NZCA 151 at [26] and *R* v Lau [2018] NZHC 2935 at [42]). And from a perception perspective, having the cases heard in general court lists by generalist judges alongside offending more commonly thought of as criminal reinforces the message that harming the environment is no different to (directly) harming people or property, in that offending of this nature will be dealt with in the same way.

This is, it is suggested, the crux of the matter. While it has been observed that "[g]reen criminology is premised on the idea that the justice system needs to take environmental harm seriously" (White 2013a, 21), this book suggests that the matter is a two-way street. For the justice system to take environmental harm seriously then environmental harm must also take the justice system, and in particular the criminal law, seriously. This goes not only to the decision as to what conduct should be criminalised, but also how the offences are drafted, what penalties are provided and what the institutional arrangements for prosecuting are.

This idea of understanding and respecting the criminal law is something that is often missing in the study of environmental non-compliance (and indeed regulatory breaches more generally). Too often the answer to the question of what should be done with those that harm the environment is that the criminal law should be used "more". There need to be more offences. More actors should be held liable, for instance by reducing *mens rea* requirements or increasing the scope of vicarious liability. There should be more (higher) penalties imposed on those found guilty. But this book suggests that the answer is actually less, but better, criminal law.

Finally, it is hoped that this book has shown the merit of a decidedly legalist angle to the socio-legal study of environmental crime. One needs to understand the criminal law—both in theory and in practice—to be able to see why seemingly fail-proof initiatives such as increasing the maximum penalty of offences does not play out as intended. A legalist approach is sometimes criticised for failing to address the "real" problems—that is, by arguing over how to improve enforcement we lose sight of the much bigger problem of defining what should be unlawful in the first place. There is, of course, something to be said for this. But what this book has shown is that there is work to be done at the enforcement level, and such work will have "real" implications should there ever be change in what is currently lawful. Put another way, there is no point winning the criminalisation battle only to see it undermined in application.

References

Ashworth, Andrew. 2010. *Sentencing and Criminal Justice*, 5th ed. Cambridge: Cambridge University Press.

Ashworth, Andrew, and Lucia Zedner. 2012. Prevention and Criminalization: Justifications and Limits. *New Criminal Law Review: An International and Interdisciplinary Journal* 15 (4): 542–571. https://doi.org/10.1525/nclr.2012.15.4.542.

Brown, Marie. 2017. *Last Line of Defence: Compliance, Monitoring and Enforcement of New Zealand's Environmental Law*. Auckland: Environmental Defence Society.

Environment Canterbury. 2021. Infringement Notices. https://www.ecan.govt.nz/your-region/your-environment/monitoring-and-compliance/investigations/infringement-notices/. Accessed 24 August 2021.

Faure, Michael, and Katarina Svatikova. 2012. Criminal or Administrative Law to Protect the Environment? Evidence from Western Europe. *Journal of Environmental Law* 24 (2): 253–286. https://doi.org/10.1093/jel/eqs005.

Ministry for the Environment. 2016. National Monitoring System for 2014/15.

Ministry for the Environment. 2017. National Monitoring System for 2015/16.

Ministry for the Environment. 2018a. Best Practice Guidelines for Compliance, Monitoring and Enforcement under the Resource Management Act 1991.

Ministry for the Environment. 2018b. National Monitoring System for 2016/17.

Ministry for the Environment. 2019. National Monitoring System—Information Requirements: Guidance for the 2019/20 Financial Year. Wellington: Ministry for the Environment.

Nurse, Angus. 2015. *Policing Wildlife: Perspectives on the Enforcement of Wildlife Legislation*. Basingstoke: Palgrave Macmillan.

O'Hear, Michael. 2004. Sentencing the Green-Collar Offender: Punishment, Culpability, and Environmental Crime. *The Journal of Criminal Law and Criminology* 95 (1): 133–276.

Preston, Brian. 2007. Principled Sentencing for Environmental Offences—Part 1: Purposes of Sentencing. *Criminal Law Journal* 31: 91–100.

Roberts, Julian. 2003. Sentencing Reform in New Zealand: An Analysis of the Sentencing Act 2002. *Australian & New Zealand Journal of Criminology* 36 (3): 249–271. https://doi.org/10.1375/acri.36.3.249.

Welsh, Lucy, Layla Skinns, and Andrew Sanders. 2021. *Sanders and Young's Criminal Justice*, 5th ed. Oxford: Oxford University Press.

White, Rob. 2010. Prosecution and Sentencing in Relation to Environmental Crime: Recent Socio-Legal Developments. *Crime, Law and Social Change* 53: 365–381.
White, Rob. 2013a. The Conceptual Contours of Green Criminology. In *Emerging Issues in Green Criminology: Exploring Power, Justice and Harm*, ed. Reece Walters, Diane Westerhuis, and Tanya Wyatt, 17–33. Basingstoke: Palgrave Macmillan.
White, Rob. 2013b. Environmental Crime and Problem-Solving Courts. *Crime, Law and Social Change* 59 (3): 267–278. https://doi.org/10.1007/s10611-013-9414-5.

Index

M. Wright, *Responding to Environmental Crimes*, Palgrave Studies in Green Criminology, https://doi.org/10.1007/978-3-030-89250-0

R

The manufacturer's authorised representative in the EU is Springer Nature Customer Service Centre GmbH, Europaplatz 3, 69115 Heidelberg, Germany. If you have any concerns regarding our products, please contact ProductSafety@springernature.com

Printed and bound by CPI Group (UK) Ltd, Croydon, CR0 4YY
30/09/2026
02225490-0004